WORKING WITH THE ANTHROPOLOGICAL THEORY OF THE DIDACTIC IN MATHEMATICS EDUCATION

This book presents the main research veins developed within the framework of the anthropological theory of the didactic (ATD), a paradigm that originated in French didactics of mathematics. While a great number of publications on ATD are available in French and Spanish, *Working with the Anthropological Theory of the Didactic in Mathematics Education* is the first directed at English-speaking international audiences.

Written and edited by leading researchers in ATD, the book covers all aspects of ATD theory and practice, including teaching applications. The chapters feature the most relevant and recent investigations presented at the 6th international conference on the ATD, offering a unique opportunity for an international audience interested in the study of mathematics teaching and learning to keep in touch with advances in educational research. The book is divided into four sections and the contributions explore key topics such as:

- The core concept of 'praxeology', including its development and functionalities.
- The need for new teaching praxeologies in the paradigm of questioning the world.
- The impact of ATD on the teaching profession and the education of teachers.

This is the second volume in the *New Perspectives on Research in Mathematics Education* series. This comprehensive casebook is an indispensable resource for researchers, teachers and graduate students around the world.

Marianna Bosch is Professor at Universitat Ramon Llull, Spain. She is an experienced researcher in ATD with a long involvement in the dissemination of research to the international audience.

Yves Chevallard is Professor Emeritus at Aix-Marseille Université, France and the main developer of the ATD.

Francisco Javier García is a researcher and lecturer at Universidad de Jaén, Spain.

John Monaghan is Emeritus Professor at the University of Leeds, UK, and Professor at the University of Agder, Norway.

NEW PERSPECTIVES ON RESEARCH IN MATHEMATICS EDUCATION – ERME SERIES

Editors of the ERME Series

Viviane Durand-Guerrier (France)
Konrad Krainer   (Austria)
Susanne Prediger   (Germany)
Nad'a Vondrová   (Czech Republic)

International Advisory Board of the ERME Series

Marcelo Borba   (Brazil)
Fou-Lai Lin   (Taiwan)
Merrilyn Goos   (Australia and Ireland)
Barbara Jaworski   (Europe, United Kingdom)
Chris Rasmussen   (United States of America)
Anna Sierpinska   (Canada)

ERME, the European Society for Research in Mathematics Education, is a growing society of about 900 researchers from all over Europe and beyond. In the ERME community and beyond, a growing body of substantial research on mathematics education raises which is shaped by the ERME spirit of communication, cooperation, and collaboration.

The contributions in the ERME Series seek to understand and improve learning and teaching of mathematics at schools, colleges and universities, as well as in informal settings (e.g., related to street mathematics or to self-organized networks of teachers). The ERME Series puts an emphasis on reflecting the institutional, societal, and cultural contexts of learners, teachers and researchers and how this context shapes research and adopts a variety of perspectives on its research field.

The volumes are written by and for European researchers, but also by and for researchers from all over the world. An international advisory board guarantees that ERME stays globally connected. A rigorous and constructive review procedure guarantees a high quality of the series.

The series aims to provide a range of books – from individual research monographs and edited collections to textbooks and supplemental reading for scholars, researchers, policy analysts and students.

Developing Research in Mathematics Education
Twenty Years of Communication, Cooperation and Collaboration in Europe
Edited by Tommy Dreyfus, Michele Artigue, Despina Potari, Susanne Prediger, Kenneth Ruthven

Working with the Anthropological Theory of the Didactic in Mathematics Education
A comprehensive casebook
Edited by Marianna Bosch, Yves Chevallard, Francisco Javier García, John Monaghan

For more information about this series, please visit: https://www.routledge.com/European-Research-in-Mathematics-Education/book-series/ERME.

WORKING WITH THE ANTHROPOLOGICAL THEORY OF THE DIDACTIC IN MATHEMATICS EDUCATION

A comprehensive casebook

Edited by Marianna Bosch, Yves Chevallard, Francisco Javier García and John Monaghan

Routledge
Taylor & Francis Group

LONDON AND NEW YORK

First published 2020
by Routledge
2 Park Square, Milton Park, Abingdon, Oxon OX14 4RN

and by Routledge
52 Vanderbilt Avenue, New York, NY 10017

Routledge is an imprint of the Taylor & Francis Group, an informa business

British Library Cataloguing-in-Publication Data
A catalogue record for this book is available from the British Library

Library of Congress Cataloging-in-Publication Data
Names: Bosch, Marianna, editor.
Title: Working with the anthropological theory of the didactic in mathematics education : a comprehensive casebook / edited by Marianna Bosch, Yves Chevallard, F. Javier García, John Monaghan.
Description: Abingdon, Oxon ; New York, NY : Routledge, 2020. | Series: New perspectives on research in mathematics | Includes bibliographical references and index.
Identifiers: LCCN 2019023790 (print) | LCCN 2019023791 (ebook) | ISBN 9780367187712 (hardback) | ISBN 9780367187705 (paperback) | ISBN 9780429198168 (ebook)
Subjects: LCSH: Mathematics–Study and teaching–Research–Congresses.
Classification: LCC QA11.2 .W675 2020 (print) | LCC QA11.2 (ebook) | DDC 510.71–dc23
LC record available at https://lccn.loc.gov/2019023790
LC ebook record available at https://lccn.loc.gov/2019023791

ISBN: 978-0-367-18771-2 (hbk)
ISBN: 978-0-367-18770-5 (pbk)
ISBN: 978-0-429-19816-8 (ebk)

Typeset in Bembo
by Taylor & Francis Books

Printed and bound by CPI Group (UK) Ltd, Croydon, CR0 4YY

CONTENTS

ILLUSTRATIONS

Figures

Tables

Boxes

CONTRIBUTORS

Marianne Achiam, Department of Science Education, University of Copenhagen, Denmark

Michèle Artaud, ADEF, Aix-Marseille Université, France

Berta Barquero, Faculty of Education, Universitat de Barcelona, Spain

Karine Bernad, Doctor in Sciences of Education, Université d'Aix-Marseille, France

Annie Bessot, Laboratoire Informatique de Grenoble, Université Grenoble Alpes, France

Marianna Bosch, IQS School of Management, Universitat Ramon Llull, Spain

Jean-Pierre Bourgade, ADEF, Aix-Marseille Université, France

Dolores Carrillo Gallego, Dpto. de Didáctica de las Ciencias Matemáticas y Sociales, Universidad de Murcia, Spain

Corine Castela, LDAR member Emeritus, Universités de Rouen, Paris Diderot, Paris-Est Créteil, Artois et Cergy Pontoise, France

Hamid Chaachoua, Laboratoire Informatique de Grenoble, Université Grenoble Alpes, France

Yves Chevallard, Professor Emeritus, Aix-Marseille Université, France

Eva Cid, Departamento de Matemáticas, Universidad de Zaragoza, Spain

Gisèle Cirade, UMR EFTS, ESPE Toulouse Midi-Pyrénées, Université Toulouse - Jean Jaurès, Université de Toulouse, France

Anne Crumière, UMR EFTS, ESPE Toulouse Midi-Pyrénées, Université Toulouse - Jean Jaurès, Université de Toulouse, France

Ignasi Florensa, Escola Universitària Salesiana de Sarrià, UAB, Universitat Autònoma de Barcelona, Spain

Cecilio Fonseca, Departamento de Matemática Aplicada I, Universidad de Vigo, Spain

Francisco Javier García, Departamento de Didáctica de las Ciencias, Universidada de Jaén, Spain

Josep Gascón, Departament de Matemàtiques, Universitat Autònoma de Barcelona, Spain

Hiroaki Hamanaka, Department of Education in Mathematics and Natural Sciences, Hyogo University of Teacher Education, Japan

Mercedes Hidalgo, Departamento de Didáctica de Ciencias Experimentales, Sociales y Matemáticas, Universidad Complutense de Madrid, Spain

Britta Jessen, Department of Science Education, University of Copenhagen, Denmark

Caroline Ladage, EA4671 ADEF, Aix-Marseille Université, France

Catarina Lucas, Institute of Public Health (ISPUP), University of Porto, Portugal

Martha Marandino, Faculdade de Educação, Universidade de São Paulo, Brasil

Yves Matheron, Institut de Mathématiques de Marseille, UMR 7373, Institut Français de l'Éducation, France

Takeshi Miyakawa, School of Education, Waseda University, Japan

Tatsuya Mizoguchi, Department of Education, Tottori University, Japan

John Monaghan, Leeds University and Agder University, United Kingdom and Norway

José M. Muñoz-Escolano, Departamento de Matemáticas, Universidad de Zaragoza, Spain

Pedro Nicolás, Dpto. de Didáctica de las Ciencias Matemáticas y Sociales, Universidad de Murcia, Spain

Kaj Østergaard, VIA University College, Denmark

Koji Otaki, Department of Teachers Training, Hokkaido University of Education, Japan

André Pressiat, Professor Emeritus, Université Paris Diderot, France

Klaus Rasmussen, University of Copenhagen & University College Copenhagen, Denmark

Esther Rodríguez, Departamento de Investigación y Psicología en Educación, Universidad Complutense de Madrid, Spain

Avenilde Romo, CICATA, Instituto Politécnico Nacional, Mexico

Noemí Ruiz-Munzón, ESCSE Tecnocampus , Universitat Pompeu Fabra, Spain

Alicia Ruiz-Olarría, Departamento de Didácticas Específicas, Universidad Autónoma de Madrid, Spain

Encarna Sánchez-Jiménez, Dpto. de Didáctica de las Ciencias Matemáticas y Sociales, Universidad de Murcia, Spain

Maggy Schneider, Institut de mathématique, Université de Liège, Belgium

Yusuke Shinno, Department of Mathematics Education, Osaka Kyoiku University, Japan

Tomás Ángel Sierra, Departamento de Didáctica de Ciencias Experimentales, Sociales y Matemáticas, Universidad Complutense de Madrid, Spain

Carl Winsløw, Department of Science Education, University of Copenhagen, Denmark

INTRODUCTION: AN INVITATION TO THE ATD

The anthropological theory of the didactic (ATD) was the name Yves Chevallard gave to the research framework in the didactics of mathematics developed since the 1980s with the first investigations of didactic transposition (Chevallard, 1985/1991). This framework was born in the project of a "science of the didactic" proposed by Guy Brousseau and initiated by his theory of didactic situations (TDS) (Brousseau, 1997). The filiation of the ATD towards the TDS is doubtless, as well as the relations to the theory of conceptual fields due to Gérard Vergnaud (1990). What these three research frames have in common is their explicit aim to experimentally model the knowledge and know-how that is at the core of teaching and learning processes. This is why Josep Gascón (1993) proposed speaking of the "epistemological paradigm" in didactics of mathematics, to distinguish it from what was the prevailing paradigm at this period, more focused on the study of learners'—and lately teachers'—cognitive processes.

This concise chapter is an introduction—or better, an invitation, in the style of Peter L. Berger (1963)—to the ATD, its methods and conceptualisations, the problems it raises and intends to approach, and also its main assumptions and foundations. It is certainly wise to start with a warning, in case the reader is not aware of it. The conceptual framework developed by the ATD is broad and not always intuitive. This is the consequence of a main methodological principle the ATD shares with all social sciences, the *emancipatory principle*, which consists in avoiding taking for granted the elements of the social world we are studying, a world we know very well because of our experience as students and citizens and sometimes also as teachers, educators or parents. Trying to control one's assumptions is a basic gesture in scientific work. The strategy proposed by the ATD consists in setting forth a wide set of basic notions used to model—or conceptually reconstruct—the didactic world in a "fresh" perspective, to avoid being contaminated by the visions of the persons and institutions that are part of this world.

Using a specific terminology and specific notations to put this terminology at work might be initially disturbing for the neophytes. However, this very disturbance is part of the implicit assumptions—or prejudices—one has when considering a research framework about education. Our culture tells us that problems about education have to be solved in an uncomplicated way, mostly using common sense and common means and perhaps some results from other research areas such as neurology, psychology, sociology, etc. The emancipatory principle mentioned above leads to a rejection of this assumption. Moreover, we will add a second methodological principle, essential to scientific creation, that we will call the *Humpty Dumpty principle*, in reference to Lewis Carroll's well-known character in chapter VI of *Through the looking-glass, and what Alice found there* (1871):

"I don't know what you mean by 'glory'," Alice said.

Humpty Dumpty smiled contemptuously. "Of course you don't—till I tell you. I meant 'there's a nice knock-down argument for you!'"

"But 'glory' doesn't mean 'a nice knock-down argument'," Alice objected.

"When I use a word," Humpty Dumpty said, in rather a scornful tone, "it means just what *I* choose it to mean—neither more nor less."

The Humpty Dumpty principle seems obvious when we consider well-established disciplines. The theorisation of reality, essential to put the emancipatory principle at work, needs new terms. This is done by inventing new words—like "heteroskedasticity", "molecule" or "unconscious"—or by assigning new meanings to common words. Mathematicians, sociologists and psychologists talk about "fields" to refer to different things, none of which have grass or flowers … . Moreover, the use of written symbols is sometimes of great help in describing complex situations and imagining new cases or relations that would not easily appear through ordinary discourse. Words, expressions and discourses, as well as symbols, graphs, figures and gestures, are crucial cognitive tools. They are indispensable for describing and scrutinising the objects and phenomena we wish to study, as well as imagining and creating new realities, new entities and new relations; talking about what exists but also about what *could* exist, but does not, as well as about its (non-existent) conditions of possibility.

Research in the ATD has generated its own system of terms, principles and relations, as well as its own way of questioning reality and approaching it—its own methodology. A third important principle in this respect is the one that gives its name to the theory: the *anthropological* one. Why is the ATD an *anthropological* theory? Why was this term chosen? What does it stand for and what does it imply? The anthropological principle affects the level of generality and specificity that is assigned to didactic phenomena. In the ATD, didactic phenomena are considered as inherent to any group of human beings, as part of humanity. Being human beings means co-creating and disseminating knowledge, and also failing to do so. Moreover, the anthropological approach also proposes a common vision of all kind of human exchange or activity, trying to exclude all kinds of value or categorisation brought about by the society and institutions researchers are immersed in—and subjected to. For instance, an important distinction when analysing teaching and learning processes affects the "content" of the process, what is taught and learnt. It can be just a basic gesture—"Show me how to open this tap"—or a body of knowledge like mathematics, music, grammar, history or economics. In between there are multiple other options, such as sawing, playing football, investing one's savings, preparing a couscous or writing a formal letter of complaint. To approach all these processes in a unitary perspective and, especially, to prevent assuming the complex distinctions our different cultures put on all these contents, the ATD proposes to initially conceptualise all of them as a same entity by using an invented word: praxeology. The anthropological principle states that any human activity can be described as a praxeology or as an amalgamation of praxeologies of different "sizes".

The term praxeology results from the union of two Greek words: *praxis* and *logos*. The praxis refers to the practical part of the activity, the "doing" and its related know-how. It is split into two elements: a type of tasks T_i and a technique τ_i to carry out the tasks of type T_i. In accordance with the Humpty Dumpty principle, tasks and techniques must be understood in a broad sense: creating a new molecule, driving a boat, encouraging a sports team or drinking water are types of tasks, and they have different possible associated techniques, not necessarily algorithmic. The second element of the praxeology, the *logos*, also contains two types of elements: a technology and a theory. If the praxis corresponds to the "doing", the logos corresponds to the "telling", the description, presentation, explanation, justification of the technique, together with the organisation of the types of tasks and their relations to the techniques. A first level of description and justification is the level of the technologies θ_i. The word is understood here according to its etymology: a discourse (*logos*) about the technique (*technè*). Technologies rely on principles and draw on concepts and terms that belong to the second level of justification, the theories Θ_i.

The anthropological principle states that any human activity can be described in terms of praxeologies. It contains a subprinciple derived from the very notion of praxeology: any human activity includes a *praxis* component or know-how together with a *logos* component of knowledge. In other words, any practice is described and justified in one way or another, sometimes very poorly—"I drink water like this because this is the way to drink water. Is there another one?"—, other times with complex developments, as the example of the creation of new molecules lets us guess. Some praxeologies can be considered as having an underdeveloped praxis because the techniques are not "performant" enough or because new types of tasks have appeared without a good way to carry them out. Other praxeologies can be considered as having an underdeveloped logos: techniques almost without any justification, without specific words to describe them, etc. And there are also praxeologies considered as having an overdeveloped logos, with too much to say for such a little doing.

Praxeologies are entities that evolve over time. New types of tasks appear, asking for new techniques to perform them; some new conditions can make a technique fail in some cases and require new developments; the new praxis will then lead to the expansion of the logos to describe, explain or justify the new ways of doing. Other evolutions are also possible. A new theoretical element produces new technologies—for instance, in mathematics, a new theorem derived from a new property or relation between notions—which lead to new ways of doing and the appearance of new types of tasks.

A last ATD principle should be mentioned about the consideration of praxeologies as evolving entities: their institutional relativity. We will use a simple mathematical example (taken from Bosch and Gascón, 2014) to illustrate it. Let us consider Pythagoras' theorem. At first sight, at least for somebody who is not far away from the educational system, this property of the lengths of the three sides of a right triangle is considered a technological element of a praxeology that exists in many lower secondary schools, linked to a praxis that consists in types of tasks like determining if a triangle has a right angle; or to calculate one of the lengths of a right triangle given the other two; etc. The development of this praxis and the corresponding logos can give rise to a whole body of knowledge—a broader praxeology—called trigonometry. However, in another context, say a topographic praxis of making measurements in a field, the Pythagorean theorem can just be a simple technique of drawing a right angle when other tools are not available. Its status and function change, as well as the justification discourse that is built around it—Pythagoras' theorem may be "a well-known property" and nothing else. Finally, Pythagoras' theorem can move from a technological or technical element to a theoretical one, when the space or its geometry is considered to be Euclidean or not. Therefore, a given object can be part of a type of tasks, a technique, a technological or a theoretical element depending on the praxeology we are considering and the way this praxeology exists and evolves in a given institutional setting. Not only does a praxeology change when moving from one institution to another (we do not drink water in the same way everywhere), the very elements of a praxeology can have different functions depending on the type of praxeology that is considered.

After this brief introduction to some basic principles of the ATD, we can invite the reader to cross the threshold to the other side of the mirror. Let us just add a final comment. As with other knowledge domains, when we talk about the anthropological theory of the didactic, we do not refer to a *theory* in the ATD sense, but to a set of *research praxeologies*. As is often the case in science, "theory" is used metonymically. This introductory chapter follows a traditional pattern, starting from the ATD basic principles, which properly belong to the theory (in the ATD sense). Many other theoretical and technological elements—primary notions, derived notions, relations, properties—are presented in the form of a glossary to be found right after this presentation. This glossary includes a list of current ATD terms with their definition and some short developments to relate them to other entries and explain some of their uses. Even if the ATD cannot be reduced to a list of terms, we hope it will help the reader to better approach fully fledged ATD research praxeologies.

Many other aspects of the ATD logos and praxis are presented in the list of chapters that make up this book. They include recent research studies that were presented at the 6th international conference on the ATD held in Castro Urdiales

(Spain) in 2016, which was recognised as an ERME Topic Conference. These studies illustrate some of the main research questions that are raised in the ATD, the way they are formulated and approached and some of the results obtained. In brief, they bring to the fore some research tasks and research techniques and their outcomes. Of course, this research *praxis* is just evoked, mainly through the technological discourses that describe, explain and justify the work done—sometimes in a summarised way due to space restrictions. The results obtained, the product of the *praxis*, are tested by the research community and, when they appear to be relevant and robust enough, they become part of the research *logos*. They will then become part of new research methodologies and yield important tools to formulate new research questions, thus enlarging the research *praxis*, its techniques and types of task. As any praxeological open system, the ATD evolves through complex interactions between the praxis and logos elements. In a written presentation, the logos elements may acquire a stronger presence. We hope that this limitation will not restrain the reader from accessing the whole of the ATD as a praxeological complex. This is the mission of the research studies presented in this book.

References

Berger, P. L. (1963). *Invitation to sociology: A humanistic perspective*. New York: Knopf Doubleday Publishing Group.

Brousseau, G. (1997). *Theory of didactical situations in mathematics: Didactique des mathématiques 1970–1990*. Dordrecht, The Netherlands: Kluwer.

Chevallard, Y. (1985/1991). *La transposition didactique: Du savoir savant au savoir enseigné*. Grenoble, France: La Pensée Sauvage (2nd edition 1991).

Gascón, J. (1993). Desarrollo del conocimiento matemático y análisis didáctico: del patrón Análisis-Síntesis a la génesis del lenguaje algebraico. *Recherches en Didactique des Mathématiques*, 13(3), 295–332.

Vergnaud, G. (1990). La théorie des champs conceptuels. *Recherches en Didactique des Mathématiques*, 10(2), 133–170.

Short bibliography on the ATD

Barbé, J., Bosch, M., Espinoza, L., & Gascón, J. (2005). Didactic restrictions on the teacher's practice: The case of limits of functions in Spanish high schools. *Educational Studies in Mathematics*, 59, 235–268.

Barquero, B., & Bosch, M. (2015). Didactic engineering as a research methodology: From fundamental situations to study and research paths. In A. Watson & M. Ohtani (eds), *Task design in mathematics education* (pp. 249–271). Cham, Switzerland: Springer.

Bosch, M. (2015). Doing research within the anthropological theory of the didactic: The case of school algebra. In S. Cho (ed.), *Selected regular lectures from the 12th International Congress on Mathematical Education* (pp. 51–70). Cham, Switzerland: Springer.

Bosch, M. (2018). Study and research paths: A model for inquiry. Proceedings of the International Congress of Mathematicians. Rio de Janeiro, 1–9 August (vol. 3, pp. 4001–4022).

Bosch, M., & Gascón, J. (2006). Twenty-five years of the didactic transposition. *ICMI Bulletin*, 58, 51–65.

Bosch, M., & Gascón, J. (2014). Introduction to the anthropological theory of the didactic (ATD). In A. Bikner-Ahsbahs & S. Prediger (eds), *Networking of theories as a research practice in mathematics education* (pp. 67–83). Dordrecht, The Netherlands: Springer.

Chevallard, Y. (1992). Fundamental concepts in didactics: Perspectives provided by an anthropological approach. In R. Douady & A. Mercier (eds), *Research in didactique of mathematics: Selected papers* (pp. 131–167). Grenoble, France: La Pensée Sauvage.

Chevallard, Y. (2006). Steps towards a new epistemology in mathematics education. In M. Bosch (ed.), *Proceedings of CERME 4* (pp. 21–30). Barcelona, Spain: Fundemi IQS.

Chevallard, Y. (2007). Readjusting didactics to a changing epistemology. *European Educational Research Journal* 6(2), 131–134.

Chevallard, Y. (2015). Teaching mathematics in tomorrow's society: A case for an oncoming counter paradigm. In S. J. Cho (ed.), *Proceedings of the 12th International Congress on Mathematical Education* (pp. 173–187). Springer International Publishing.

Chevallard, Y. (2019) Introducing the anthropological theory of the didactic: An attempt at a principled approach. *Hiroshima Journal of Mathematics Education*, 12, 71–114.

Chevallard, Y., & Bosch, M. (2014). Didactic transposition in mathematics education. In S. Lerman (ed.) *Encyclopedia of mathematics education* (pp. 170–174). Dordrecht, Netherlands: Springer.

Chevallard, Y., & Sensevy, G. (2014). Anthropological approaches in mathematics education: French perspectives. In S. Lerman (ed.), *Encyclopedia of mathematics education* (pp. 38–43). Dordrecht, The Netherlands: Springer.

Chevallard, Y., & Bosch, M. (2020). The anthropological theory of the didactic. In S. Lerman (ed.), *Encyclopedia of mathematics education*. Dordrecht, Netherlands: Springer.

Winsløw, C. (2015). Mathematics at university: The anthropological approach. In S. Cho (ed.), *Selected regular lectures from the 12th International Congress on Mathematical Education* (pp. 859–876). Cham, Switzerland: Springer.

A SHORT (AND SOMEWHAT SUBJECTIVE) GLOSSARY OF THE ATD

Yves Chevallard, with Marianna Bosch

> Hence it appears that definitions are very arbitrary, and that they are never subject to contradiction; for nothing is more permissible than to give to a thing which has been clearly designated, whatever name we choose. It is only necessary to take care not to abuse the liberty that we possess of imposing names, by giving the same to two different things.
>
> Blaise Pascal, *The Geometric Spirit and the Art of Persuasion*, ca. 1658

Activity.   The word applies to all stabilised or slowly evolving, not haphazard or erratic, practices in which someone engages. The main tenet of the ATD in this respect is that all human activity breaks down into a number of *tasks* of determined *types*. (See also *Types of tasks*.)

Answer.   A first type of answers responds to *how questions*: basically, these answers are made up of a *technique* which indicates how to accomplish *tasks* of a certain *type*, to which is added the necessary *technological* and *theoretical block* in order to get a complete *praxeology*. A second type of answers responds to *why questions*: they usually consist in a *technological* and *theoretical block*, which may generate new *praxis blocks*, or come to play the *logos* part in a number of praxeologies in need of a *logos* part. (See also *Questions*.)

Antididactic.   See *Didactic gestures and the possibly didactic*.

Break.   A break is a change in a person's relations to a number of objects such that the person's new relations do not conform to the corresponding relations of any surrounding institution, thus laying the basis for a potential, conspicuously new institution. Although such a change should be called generally a *cognitive break*, it is traditionally labelled an *epistemological break*. A rupture from a given institution's relations to a number of objects is termed an *institutional break*.

Civilisation. In the ATD, this word is not taken to mean a process—the process of civilisation. It applies to a set of societies that, up to a point, can be said to share—through historical fashioning—a large set of *objects* and *relations* to these objects. The concept is functional in pointing to conditions and constraints present in a whole range of societies, albeit under a specific guise. It helps in avoiding false beliefs about conditions at first regarded as local—and therefore "shallow"—when in fact they are deeply rooted in a whole civilisation—and therefore not so easily removable. (See also *Scale of didactic codeterminacy levels*.)

Cognition. Persons and institutions are *objects*. In the ATD, an object is anything that exists for (at least) a person x or an institutional position p. Given a person x and an object o, the *personal relation* of x to o, denoted by $R(x, o)$, the set of links of any kind between x and o. It may be that $R(x, o) = \varnothing$. In this case, we say that "o does not exist for the person x" or that "x does not know the object o". If, on the contrary, $R(x, o) \neq \varnothing$, the object o exists for x, and x knows o, even if in a minimalist way. The same definitions apply when replacing the person x with an institutional position (I, p). The relation of p to o is traditionally denoted by $R_I(p, o)$. We have either $R_I(p, o) = \varnothing$ or $R_I(p, o) \neq \varnothing$. We say that the institutional position p does not know or on the contrary, knows the object o, and as well that o does not exist or exists for p.

Cognitive algebra. The word "algebra" is taken here slightly metaphorically. When an author writes "We have" or "One has", as in "We have either $R_I(p, o) = \varnothing$, or $R_I(p, o) \neq \varnothing$", who is the "one" or the "we"? More specifically, when the author writes "We have $R(\hat{\imath}, o) \neq \varnothing$", the "we" is necessarily some personal or institutional instance $\hat{\jmath}$ (of course $\hat{\jmath}$ may be the author). This is denoted by $\hat{\jmath} \vdash R(\hat{\imath}, o) \neq \varnothing$, and, more generally, by $\hat{\jmath} \vdash \vartheta$, where ϑ is any sentence. The statement $\hat{\jmath} \vdash \vartheta$ is to be read "the instance $\hat{\jmath}$ judges that ϑ [is true]". (The symbol $\vdash$ used here is the Unicode "assertion character".) Another instance $\hat{k}$ may judge that ϑ is not true, i.e., $\hat{k} \vdash \neg \vartheta$. There is not generally agreement but disagreement between different instances. This is an aspect of *cognitive relativity*—as a general rule, instances "do not think alike". All this can be easily generalised. The statement $\hat{k} \vdash (\hat{\jmath} \vdash R(\hat{\imath}, o) \neq \varnothing)$ means, for example, that, according to the instance $\hat{k}$, the instance $\hat{\jmath}$ thinks that the $\hat{\imath}$ knows the object o. Likewise, if a person x_0 asserts "They say that the person x_1 knows the person x_2", we shall write: $x_0 \vdash \exists \hat{\imath} \, (\hat{\imath} \vdash R(x_1, x_2) \neq \varnothing)$, where the instance $\hat{\imath}$ stands for "they". The fact that $\hat{\jmath} \vdash R(\hat{\imath}, o) \neq \varnothing$ (or, as well, that $\hat{\jmath} \vdash R(\hat{\imath}, o) = \varnothing$) tells us something about the relation $R(\hat{\jmath}, R(\hat{\imath}, o))$, i.e., about the relation of $\hat{\jmath}$ to the relation of $\hat{\imath}$ to o. The fact that $\hat{k} \vdash (\hat{\jmath} \vdash R(\hat{\imath}, o) \neq \varnothing)$ informs us about $R(\hat{k}, R(\hat{\jmath}, R(\hat{\imath}, o)))$, i.e., about the relation of $\hat{k}$ to the relation of $\hat{\jmath}$ to the relation of $\hat{\imath}$ to o, etc.

Cognitive base. See *Cognitive nucleus*.

Cognitive break. See *Break*.

Cognitive equipment.   Given an instance $\hat{\imath}$, the universe of objects or *cognitive universe* of $\hat{\imath}$ is defined by: $\Omega(\hat{\imath}) \triangleq \{o/R(\hat{\imath}, o) \neq \varnothing\}$. The set $\Omega(\hat{\imath})$ tells us which objects exist for $\hat{\imath}$. The *cognitive equipment* of $\hat{\imath}$ is then defined by: $\Gamma(\hat{\imath}) \triangleq \{(o, R(\hat{\imath}, o))/o \in \Omega (\hat{\imath})\}$. The set $\Gamma(\hat{\imath})$ specifies how $\hat{\imath}$ knows the objects they know, i.e., the objects $o \in \Omega(\hat{\imath})$. (See also *Praxeologies and praxeological analysis*.)

Cognitive nucleus.   Let $\hat{w}$ be the reference instance. Let t_0 and t_1 be two successive times. How can we make sense of the following semi-formalised statement $\hat{w}$ ⊢ [$R(\hat{\imath}, o)$ at t_1 is "better" than $R(\hat{\imath}, o)$ at t_0]? At the start, $\hat{w}$ has in mind the ordered pair $\bar{n} = (\hat{\imath}, o)$, called the *cognitive base*. This base will be completed so that $\hat{w}$ can arrive at a judgment. To do so, we must assume that $\hat{w}$ also envisages a positional instance (I, p) whose relation to o, $R_I(p, o)$, is regarded by $\hat{w}$ as a reference point against which $R(\hat{\imath}, o)$ can be evaluated: the more $R(\hat{\imath}, o)$ will be recognized as "close" to $R_I(p, o)$ and the better it will appear. But this is not enough. The reference instance $\hat{w}$ must also consider an *evaluating instance* $\hat{v}$, which is supposed to judge the "proximity" of $R(\hat{\imath}, o)$ to $R_I(p, o)$. The triple $\underline{n} = (I, p, \hat{v})$ is the *cognitive reference frame*. If $\hat{w}$, i.e., the reference instance, is a teacher y, the relation $R_I(p, o)$ may be the one expected at the exam for which y is preparing the *students* $x = \hat{\imath}$, the evaluating instance $\hat{v}$ being the examination board. The quintuple $\tilde{n} = \overbrace{\bar{n} \; \underline{n}} = (\hat{\imath}, o, I, p, \hat{v})$ is a *cognitive nucleus* (or kernel). The instances (I, p), $\hat{v}$, and even $\hat{\imath}$, are often merely imagined by $\hat{w}$ and are therefore respectively denoted by $^*I$, $^*p$, $^*\hat{v}$ or $^*\hat{\imath}$. The corresponding cognitive nucleus, $^*\tilde{n}$, is written as the case may be, $(\hat{\imath}, o, ^*I, ^*p, ^*\hat{v})$, $(\hat{\imath}, o, ^*I, ^*p, \hat{v})$, $(\hat{\imath}, o, I, ^*p, ^*\hat{v})$, etc. Cognitive nuclei are therefore frequently underdefined. Note that, of course, we can have, for example, $\hat{v} = \hat{w}$ or else $\hat{v} = \hat{\imath}$. The evaluating instance $\hat{v}$ is supposed to be able to pronounce judgments like "$R(\hat{\imath}, o)$ conforms with $R_I(p, o)$" or "does not conform to $R_I(p, o)$", which are to be denoted, respectively, by $R(\hat{\imath}, o) \cong R_I(p, o)$ and $R(\hat{\imath}, o) \ncong R_I(p, o)$. But it is also assumed that $\hat{v}$ is able to estimate a "degree of conformity" $\varphi(R, \bar{R})$ between $R = R(\hat{\imath}, o)$ and $\bar{R} = R_I(p, o)$. Given relations R and R' to o, we usually have one of the three following verdicts: $\hat{v} \vdash \varphi(R, \bar{R}) < \varphi(R', \bar{R})$; $\hat{v} \vdash \varphi(R, \bar{R}) > \varphi(R', \bar{R})$; $\hat{v} \vdash \varphi(R, \bar{R}) \approx \varphi(R', \bar{R})$. In particular, R and R' can be respectively the relation $R(\hat{\imath}, o)$ at times t and t', with $t < t'$. The notion of cognitive nucleus is the notion on which hinges the decisive concept of a *possibly didactic situation*.

Cognitive reference frame.   See *cognitive nucleus*.

Cognitive relativity.   See *cognitive algebra*.

Cognitive universe.   See *cognitive equipment*.

Conditions and constraints.   In the ATD, a *condition* is anything purported to have influence over at least something. Essentially, a condition can thus be identified to the state of some system, and therefore to the values of a set of variables that govern this system. A *constraint* is any condition which appears to be unmodifiable by occupants—acting as such—of a given institutional *position*. Very often, for any condition, there exists at least one position in society from which this condition can be modified. It is the universal aim of the natural as well as the social and behavioural sciences to create means to turn ancient, lasting constraints into conditions modifiable from at least some (in general newly created) institutional positions.

Constraint.   See *Conditions and contraints*.

Curricular path.   See *Curriculum*.

Curricular trail.   See *Curriculum*.

Curriculum.   The word *curriculum* derives from Latin *currere* "to run" and has come to mean a "course of study" at a college, university, or school. In the ATD it refers to a general, nonnormative notion. We say that, as seen from the instance $\hat{w}$'s vantage point, an institutional position p_s is an antecedent of a position p if p_s is a position seen by $\hat{w}$ as preparing persons x to occupy the position p, which can be denoted by $\hat{w} \vdash p_s \rightsquigarrow p$. The position p_s may be regarded, or not, as a *student* position by some "authorised" institution such as a university. We define a $\hat{w}$-curricular path from p_1 to $p = p_n$ to be a (finite) sequence of positions $p_1, p_2, \ldots, p_n = p$ so that $\hat{w} \vdash p_i \rightsquigarrow p_{i+1}$ for $1 \le i \le n - 1$. To the path $(p_1, p_2, \ldots, p_n)$ corresponds the *curricular trail* $(\pi(p_1), \pi(p_2), \ldots, \pi(p_n))$, where $\pi(p_i)$ is the *praxeological equipment* of position p_i. For the sake of brevity, we leave implicit, for each position p_i of the positional path $(p_1, p_2, \ldots, p_n)$, the potential position $\bar{p}_i$ of "teacher" (in an extended sense of the word) that may exist according to $\hat{w}$. We then define a $\hat{w}$-*curriculum* $(\pi_1, \pi_2, \ldots, \pi_n)$, where $\pi_i = \pi(p_i)$, for $1 \le i \le n$, as the curricular trail associated with the $\hat{w}$-positional path $(p_1, p_2, \ldots, p_n)$. Note that a curriculum is an existing social reality (as seen by an instance $\hat{w}$) and must not be mistaken for a curricular project. The consideration of the notion of curriculum reveals many key questions. Given a $\hat{w}$-curricular path $(p_1, p_2, \ldots, p_n)$, what percentage of people in position $p = p_n$ have arrived at this position by following the path $(p_1, p_2, \ldots, p_n)$? More generally, what percentage of those in position p_i have arrived there by following the path $(p_1, p_2, \ldots, p_i)$? For $1 \le i \le n - 1$, what percentage of "students" in position p_i gain access to the position p_{i+1}? How can a person accede to the initial position p_1? Does $\hat{w}$ acknowledge other curricular paths $(p'_1, p'_2, \ldots, p'_m)$, with $p'_m = p$? There also exist a whole array of questions about what goes on in each and every position p_i and in particular about the praxeological equipement $\pi(p_i)$ or about the links between $\pi(p_i)$ and $\pi(p_j)$ with $j > i$, including the praxeological obstacles to the establishment of $\pi(p_j)$ that $\pi(p_i)$ may generate. All this pertains to the crucial question of *curriculum development*.

Curriculum development.   See *Curriculum*.

Deranging conditions. See *Didactic gestures and the possibly didactic.*

Derived questions. See *Herbartian schema.*

Dialectic. Any *praxeology* that enables one to overcome two opposed types of *constraints* by turning them into a new kind of *conditions* that supersede them. In this context, one, therefore, speaks of *supersession* (French *dépassement*, German *Aufhebung*, Spanish *superación*).

Dialectic of black boxes and clear boxes. Praxeology that allows one, when confronted to some praxeological element, to manage one's way between full ignorance (black box) and supposedly complete knowledge (clear or white box) of that element. To cast it in formulaic style: this dialectic helps one determine the right "shade of grey" to work with— there is no such thing as a purely "white" box.

Dialectic of conjecture and proof (or dialectic of media and milieus). In the course of an inquiry on a question Q by a didactic system $S(X; Y; Q)$, X is confronted with statements expressed by what is generically called *media*, a *medium* being any system that issues messages—a textbook, a teacher, a newspaper, the Internet are all media. Of course, this list should also include Y as well as X insofar as this group utters statements regarding the question Q. Notwithstanding their plausibility, mostly all the statements "received" by X (including those coming from Y) should be regarded as conjectures, i.e. as statements based on incomplete evidence. Looking for evidence is thus the sinews of inquiry. Proof of a statement ϑ should be looked for by questioning media which, with respect to ϑ, behave like "adidactic" *milieus*. Such an adidactic milieu—or simply milieu, if no ambiguity is to be feared—is a system deemed to be devoid of any intention to prove or disprove σ, much like a part of the inanimate world. The dialectic of media and milieus enables the pursuit of truth—even in cases where there is no decisive test.

Dialectic of dissemination and reception. Whatever the answer $A^{\heartsuit}$ to some question Q, be sure that it will disseminate outside of $S(X; Y; Q)$. For example, if $[X; Y]$ is a school class, $A^{\heartsuit}$ will be known, in essence, to other teachers, to parents, etc. Bringing an answer to a question is a social act, the product of which cannot be restricted to a single place—"leaks" are sure to happen. The dissemination that takes place alters the ecology of $A^{\heartsuit}$ and may, therefore, diminish its viability, even within $[X; Y]$. How $A^{\heartsuit}$ will be received is thus a crucial concern for its producers and potential users. The *dialectic of dissemination and reception* is, therefore, a key tool of inquiry.

Dialectic of media and milieus. See *Dialectic of conjecture and proof.*

Dialectic of on-topic and off-topic. At school, the course followed by an inquiry is traditionally supposed to remain on-topic all the time, without wandering off-topic even for a short detour that would seem promising in terms of unexpected but hopefully relevant encounters. Proper mastery of the dialectic of on-topic and off-topic makes it possible to overcome this institutional limitation and go away at times from the apparent right path, in search of the unforeseen.

Dialectic of persons and institutions.   The tension between people and institutions and its supersession is at the heart of the continued creation of a society. For a person x, "being a member" of an institution I means occupying a certain position p: we say that the person x is *subjected* to I in position p or is a *subject of I* in position p. For example, a member of a school class may hold the position of *student* or the position of *teacher* (or such other existing position); a member of a family may hold the position of mother, father, or child; etc. A given person x is, and has been, subjected to a multiplicity of institutional positions (I, p). All persons are the results of the dynamical system of their institutional subjections, which gives them their power of thought and action. A person x can occupy at the same time the positions of adult, female, mother, teacher—a person is a plural reality (the word *person* originally meant a mask, that Greek actors used to wear, and of which they could change). When we address a person, we often question, in that person, the subject of this or that institutional position: "You who are a man…", "She is the form teacher", "You are the midwife?", etc. A person x may want to free themself from some of their institutional subjections. Such desubjections are made at the cost of contracting new subjections, which frees up old ones and gives new power of thought and action while creating new constraints. Such trade-offs are constitutive of the person x's freedom.

Dialectic of questions and answers.   See *Dialectics of inquiry* and *Herbartian schema.*

Dialectic of reading (or "excribing") and writing (or inscribing).   Most information comes to us in texts, as happens with the answers $A^\Diamond$ appearing in the developed Herbartian schema. Texts are made of assertions that both follow from and manifest praxeologies which, usually, remain hidden to the casual reader. These praxeologies have been "inscribed" (and thus concealed) in the text, so to speak; conversely, the serious reader, who feels concerned with the praxeologies put to use to produce the assertions he reads, will have to "undo" the inscribing by—to use a neologism—"excribing" them, i.e. by questioning the text about its hidden content, so as to bring to the fore normally latent praxeologies. It follows from all this that, reciprocally, in producing $A^\Diamond$, X (and therefore Y) has to devote much effort to "inscribing" it into the text that will preserve it from oblivion and make it known more widely. Altogether, all this necessitates considerable writing and above all different kinds of writing (such as in a notebook, a progress report, a draft, etc.)

Dialectic of the individual and the group (or dialectic of *idionomy* and *synnomy*).   In a school class $[X; Y]$ where $Y = \{$the teacher $y\}$, it is usually supposed that every student $x \in X$ inquires about a question Q on his or her own to produce an answer $A^\heartsuit_x$. In general, the answer $A^\heartsuit_x$ supplied by x will cease to have any relevance from the very moment the teacher discloses his or her own answer $A^\heartsuit_y$, which will displace all answers $A^\heartsuit_x$, $x \in X$, in accordance with the degenerate *Herbartian schema*: $S(X;\ Y;\ Q) \rightsquigarrow A^\heartsuit_Y$. This leads students to develop an individualistic relation to knowledge (and to ignorance), caught as they are between their *idionomy* (from Greek *idios* "one's own" and *nomos* "law") and the *heteronomy* imposed by the teacher. By contrast, in an inquiry as modeled by the (non degenerate) Herbartian schema, answers $A^\heartsuit_x$ are no more but no less than answers $A^\Diamond$ that will constitute part of the milieu M from which $A^\heartsuit$ is to be produced according to the semi-developed Herbartian schema, $[S(X;\ Y;\ Q) \rightarrow M] \rightsquigarrow A^\heartsuit$, with $M = \{A^\Diamond_1, A^\Diamond_2, ..., A^\Diamond_n, W_{n+1}, ..., W_m, ...\}$. In such a perspective, a student is no longer accountable *only* for his or her own answer $A^\heartsuit_x$: *all* students are collectively accountable for the answer $A^\heartsuit$ and its construction. Their main need is, therefore, to establish in the class a common law, determined and applied collectively, to which they will be accountable. Such a *synnomy* (from Greek *syn* "together") must counterbalance the idionomy that remains indispensable for each student in his or her personal effort to investigate the question Q and bring his or her share to the advancement of the inquiry.

Dialectic of the parachutist and the truffle hound. When looking for information in the course of some inquiry, one has to sweep vast areas, thus acting as a (military) parachutist, while knowing that the information searched for will be found (in the way a truffle hound—or hog, or pig—does) only in some sporadic, unexpected places. The capacity to do so is identical with the mastery of the *dialectic of the parachutist and the truffle hound*.

Dialectics of inquiry. In the ATD, the action taken to provide an answer A to a question Q is called an *inquiry* into Q. An inquiry is minimally represented by the so-called *reduced Herbartian schema* $S(X;\ Y;\ Q) \rightsquigarrow A^\heartsuit$, in which the exponent ♥ (heart) means that the answer $A^\heartsuit$ satisfies conditions specific to $S(X;\ Y;\ Q)$, i.e., is an answer according to $S(X;\ Y;\ Q)$'s heart, so to speak. In carrying out an inquiry, X and Y should put to use the dialectics of inquiry, namely the dialectics *of on-topic and off-topic, of the parachutist and the truffle hound, of black boxes and clear boxes, of conjecture and proof* (also called dialectic *of media and milieus*), *of reading* (or "*excribing*") *and writing* (or *inscribing*), *of dissemination and reception, of the individual and the group* (also called dialectic *of idionomy and synnomy*). All the work done generates derived questions and therefore partial answers from which the answer $A^\heartsuit$ will be produced. So that the dialectics mentioned above are carried out in constant interaction with the key *dialectic of questions and answers*.

Didactic. See *Didactic gestures and the possibly didactic*.

Didactic analysis. A didactic analysis is an analysis of the *didactic* present in a given social situation according to a given instance $\hat{w}$. Such analysis should include an account of every *didactic system* $S(X; Y; \heartsuit)$ which must consist minimally in more or less comprehensive answers to the following questions: What is X? What is Y? What are the questions or praxeologies $[T/\tau/\theta/\Theta]$ that the didactic stake $\heartsuit$ is made of? What are the didactic praxeologies put to use by X and Y and what didactic means have proved necessary to do so? How does $S(X; Y; \heartsuit)$ evolve over time? What *praxeological equipment* can be engendered in X as a short-term and as a long-term result of the functioning of $S(X; Y; \heartsuit)$? And lastly, what does Y as well as some institutional environments of $S(X; Y; \heartsuit)$ may have learnt in the process? To answer these questions properly, it seems crucial to identify the main conditions and constraints composing the *ecology* of the situation and their potential effects on it. To do so, one should then scan the *scale of levels of didactic codeterminacy* to make explicit conditions or constraints too often described—or even simply alluded to— as "natural" and ("therefore") inconsequential. Didactic analysis is carried out through the elaboration of models (in the scientific sense) that are used to provide a description of the elements of $S(X; Y; \heartsuit)$ and their evolution. In the ATD, these models pertain to the theory of *didactic moments* when $\heartsuit$ is a praxeology or an amalgamation of praxeologies, and to the *Herbartian schema* when $\heartsuit$ is a question Q formulated within a project Π.

Didactic gestures and the possibly didactic.   The word *gesture* is used in an extended and neutral sense. That being said, how then can we define a *didactic gesture*? Strictly speaking, the answer is: there's no way of doing it simply. We can only speak of *possibly didactic* gestures. But in fact every gesture is "possibly didactic". Let $\hat{w}$ be a *reference instance* and $\tilde{n} = (\hat{\imath}, o, I, p, \hat{v})$ a *cognitive nucleus*. Let us suppose that some instance $\hat{u}$ "makes a gesture" δ. When will we say that the gesture δ is *didactic for* $\hat{w}$, or $\hat{w}$-*didactic*, with respect to $\tilde{n}$? To do so, we have to introduce a pivotal parameter, the set C made up of all the *conditions* that prevail *before* the gesture δ takes place. The quadruple $\varsigma = (\tilde{n}, \hat{u}, \delta, C)$ then denotes a *possibly didactic situation*. Let R be the relation $R(\hat{\imath}, o)$ before the gesture δ is performed and let R' be the "same" relation after the accomplishment of δ. We say that the gesture δ is *didactic for* $\hat{w}$ (or $\hat{w}$-*didactic*) with respect to $\tilde{n}$ and C if $\hat{w}$ conjectures that $\hat{v}$ will consider R' "closer" to $\bar{R} = R_I(p, o)$ than was the case of R. If $\hat{w}$ conjectures that $\hat{v}$ will consider it further away from $\bar{R} = R_I(p, o)$ than was the case of R, we say that the gesture δ is *antididactic for* $\hat{w}$ (or $\hat{w}$-*antididactic*). The gesture δ is said to be *isodidactic for* $\hat{w}$ (or $\hat{w}$-*isodidactic*) with respect to $\tilde{n}$ and C if $\hat{w}$ conjectures that $\hat{v}$ will find R and R' are almost equally compliant with $\bar{R}$. To conjecture that the gesture δ (performed by $\hat{u}$) is *didactic* or *isodidactic* or *antididactic* with respect to $\tilde{n}$ and C, $\hat{w}$ relies in particular on $\hat{w}$'s relations to $R(\hat{\imath}, o)$, (I, p), $\hat{v}$ and C, i.e., on $R(\hat{w}, R(\hat{\imath}, o))$, $R(\hat{w}, (I, p))$, $R(\hat{w}, \hat{v})$ and $R(\hat{w}, C)$. The gesture δ changes the prevailing conditions C, in particular, because it modifies $R(\hat{\imath}, o)$. Let us denote by C' the new set of conditions created by δ. We call C' a *derangement* of C and write $C' = C^{\lambda\delta}$ (which can be read "C deranged by δ"), where the symbol used (λ) is the caret insertion point. We thus have $C' = C_0 \cup D_\delta$, with $C_0 \subseteq C$ and $D_\delta \cap C = \varnothing$. This may lead to rewrite the situation ς as $\hat{\varsigma} = (\tilde{n}, \hat{u}, D, C)$, where D is the set of *deranging conditions*. Note that the instance $\hat{w}$

forms their judgment before the gesture δ takes place about the judgment that the evaluating instance $\hat{v}$ will issue after δ has been performed. In the case when ς and δ are (roughly speaking) reproducible, $\hat{w}$ may have integrated into their relation to ς results observed in past occurrences of the situation. But $\hat{w}$ will nevertheless issue an *a priori* judgment relating to the *a posteriori* judgment of $\hat{v}$. One of the major difficulties of forecasting in general (and not only in didactics) is our lack of knowledge about the set C of conditions and their effects on $R(\hat{i}, o)$. The commendable effort to neutralize some of these conditions does not eliminate the fact that we are unaware of even more of them. Although the possibly didactic is always unsure, it is a vital necessity to human societies, who nonetheless tend to repress it as if it were an insuperable flaw. They therefore hide it in selected, isolated places, e.g., schools and classrooms. Didacticians should however look for the *possibly didactic* wherever it occurs in society—not only where society pretends to maintain it.

Didactic milieu.   See *Dialectic of conjecture and proof* and *Herbartian schema*.

Didactic moments.   These are the moments discernible in study processes. In the study of a *how question* relating to a type of tasks T, ATD recognizes six such "study moments": the *moment of the first encounter* with T; the *moment of exploration* of T and *emergence* of a technique τ; the moment to build the *technological and theoretical block* [θ/Θ]; the moment *to work on* the praxeology produced, [T/τ/θ/Θ] and particularly on the technique τ; the moment to *institutionalise* it; the moment to *evaluate* the praxeology produced and one's relation to it. (See also *Moment.*)

Didactic organisations.   Given an object o and an instance $\hat{i}$, how can the relation $R(\hat{i}, o)$ be formed? In the case where $\hat{i}$ is a person x, the basic answer is: because x has occupied (or occupies) the *position of student* in at least one formal or informal didactic system S(X, Y, ♥) whose functioning involves the object o, notably, but not only, if ♥ is the object o itself. In all such cases, o is the subject of an *inquiry* (in an extended sense of this word). In many cases, this inquiry is still-born, or quickly vanishes, or vegetates indefinitely. This is the basic regime of the possibly didactic: individually or collectively, one quickly "forgets" to inquire into a word, a use, an event, a phenomenon. There is atrophy and repression of the (possibly) didactic. Didactics could stop here. But *Homo sapiens*, the "learned" man, is just as much *Homo discens*, the "learning" man, and *Homo docens*, the "teaching" man. For this reason, the (possibly) didactic is everywhere around us: the economy of the didactic is flourishing and could be even much more, under well-known or new forms, as revealed by the ecology of the didactic. To study the didactic, from an economic and ecological point of view, we can start from the level of the didactic system S(X, Y, ♥). Such a system is formed, under the aegis of an institution assuming a school function, by a social contract that brings together people x ∈ X, possibly other people y ∈ Y (we can have Y = ∅), and a *didactic stake* ♥. We have to examine what S(X, Y, ♥) will do and not do to inquire into ♥. In the case of the *paradigm of visiting works*, Y contains (and often is reduced to) a distinguished element y, the teacher. The didactic stake ♥ is a work o on which y has investigated prior to the formation of $S(X, Y, ♥)$ and on which the $x \in X$ are supposed not to have investigated on their own account. Under the name of *lecture*, y then presents to X a "report of inquiry" μ_y, in which the $x \in X$ have no part and through which they are supposed to "learn" o. Here, the *topos* of y and the *topos* of the $x \in X$ are essentially disjointed. There are many variations: for example, y can choose an inquiry report μ_z of which y is not the author and then guide the $x \in X$ in their study of μ_z. In recent decades, in an increasing number of societies, the *paradigm of visiting works* has increasingly been viewed as antididactic. At the same time, we observe the rise of a paradigm called the *paradigm of questioning the world*, in which the work o is a *question* q and the inquiry into q is carried out "in the classroom" by the $x \in X$ under the supervision of y. This opens up a sometimes dissonant variety of didactic organisations. The study of any work o is generally limited to the study of questions q_1, q_2, ..., .., q_n relating to o (origin, structure, use of o, etc.). But, in general, the study of o is not generated in the classroom by the study of a question q_0, so that X will not wonder whether o could be a possible tool in the study of the generating question q_0. For this and other reasons, the pedagogies of inquiry are considered by some instances as antididactic. For example, if we look at the work o as a tool to inquire into the question q_0, we will not necessarily ask ourselves all the questions asked in older curriculums which, de facto, were held to "define" what it was like to "know o". We have here a huge field of questions, which has been worked on for many years by a growing number of researchers in the context of the ATD.

Didactic stake.   See *Didactic organisations* and *Didactic system.*

Didactic system.   A didactic system $S(X; Y; ♥)$ results from the forming, around what is called a *didactic stake*, ♥, of a group made up of two functionally distinct subgroups, X and Y, the former being composed of *students* of ♥ while the latter is the team of "study assistants" (or "helpers") among whom is usually at least one "study director" who directs (and, up to a point, plans) the study underway. The most eye-catching didactic systems are those formed at school, in classrooms; in such a case, X is the class, and Y is ordinarily a "singleton" (in mathematical parlance) whose unique member is "the teacher" y.

Didactic transposition.   See *Praxeologies and praxeological analysis.*

Didactician.   A researcher in didactics.

Didactics.   Didactics is defined as the science—still in infancy—whose object is to study the conditions and constraints that govern the dissemination of praxeologies in institutions across society. Dissemination is taken here in an extended sense: it also includes *non*-dissemination and the most notably deliberate withholding of praxeologies concerning specific institutions.

Discipline. All institutional stabilised human activity is "disciplined" in the sense that it draws on determined *praxeological equipment*: the discipline of the activity is then tantamount to using that very equipment. Of course, the praxeological changes that affect any institution entail periods in which the activity within the institution is "dedisciplined", before being "redisciplined" after a while. "Discipline" in ATD thus covers much more than the slowly evolving repertoire of "school disciplines": it applies equally well to institutional disciplines with a much lesser pedigree. The conscientious reader should also note that the word *discipline* has meant from the start the "instruction given to a disciple", a meaning which by far predates the notion of "order necessary for instruction" and that of "treatment that corrects or punishes."

Ecology and economy. Abstractly, an *ecology* is simply a set of *conditions*. The ecology of a system of a given type, whether animate or inanimate, existing in the natural or the social world is the set of prevailing *conditions* under which this system lives in actual fact. It is said to be a hostile *ecology* if it causes the system to malfunction or, a fortiori, to cease to exist. A deliberately created condition (*economy*) becomes a condition of the prevailing *ecology*. The economics of the (possibly) didactic studies the conditions which, under the set of given initial conditions C, are created by specific gestures δ. Ecology, on the other hand, studies which conditions are, under given initial conditions C, possible, difficult but possible or just impossible.

Economy. See *Ecology and economy*.

Epistemological break. See *Break*.

Evaluating instance. See *Cognitive nucleus*.

Evaluation. To evaluate something is to determine its value. However, this is only half the story: there is no such thing as an intrinsic value. The value assigned depends on the project in which the "something" is called to play a part. In the *reduced Herbartian schema* $S(X; Y; Q) \rightsquigarrow A^\heartsuit$, the value of answer $A^\heartsuit$ depends on what use will be made of it. The same is true of any praxeology or, for that matter, of a person's relation to some object: their value is dependent on the situations in which the person will have to make use of that praxeology or to relate to that object. (See also *Cognition*.)

Genre of tasks. See *Types of tasks*.

Gesture. See *Didactic gestures and the possibly didactic*.

Herbartian schema. Johann Friedrich Herbart (1776-1841) was a German philosopher and the founder of pedagogy as an academic discipline. The saying goes that, according to Boyer's law, "mathematical formulas and theorems are usually not named after their original discoverers". The Herbartian schema is no counterexample to the extended law referred to Carl B. Boyer (1906-1976), though its name is not a misnomer either, for it retains something of Herbart's pedagogical views. Minimally, the Herbartian schema requires a question Q and a group of persons X—which may be reduced to a singleton—with the project to study question Q. This induces the formation of a didactic system $S(X; Y; Q)$, where the team of "study helpers" Y may be empty ($Y = \varnothing$). The functioning of $S(X; Y; Q)$ must lead to the production of an answer $A^{\heartsuit}$ to question Q, a process represented thus: $S(X; Y; Q) \hookrightarrow A^{\heartsuit}$. This is the *reduced Herbartian schema*. To produce $A^{\heartsuit}$, however, $S(X; Y; Q)$ needs "materials"; these materials make up the *didactic milieu* M constituted by $S(X; Y; Q)$ and represented as follows in the *semi-developed* Herbartian schema: $[S(X; Y; Q) \rightarrowtail M] \hookrightarrow A^{\heartsuit}$. In the didactic milieu M, it is customary to distinguish different categories. First, M accommodates "readymade" *answers* $A_i^{\diamond}$ drawn from available "resources" (including members of $X \cup Y$) and *derived questions* Q_k induced by the study of Q and $A_i^{\diamond}$. It also includes other *works* W_j specifically drawn upon to make sense of $A_i^{\diamond}$, analyze and "deconstruct" them, bring appropriate answers to the questions Q_k, and, last but not least, build up $A^{\heartsuit}$. Finally, M includes sets of data of all natures the D_l gathered in the course of the system's inquiry on Q, on which the didactic system's answers to the questions under consideration partially rest. The *didactic milieu* is therefore represented generically thus:

$$M = \{A^{\diamond}{}_1, A^{\diamond}{}_2, \ldots, A^{\diamond}{}_m, W_{m+1}, W_{m+2}, \ldots, W_n, Q_{n+1}, Q_{n+2},$$

$$\ldots Q_p, D_{p+1}, D_{p+2}, \ldots, D_q\};$$

hence the *developed Herbartian schema*:

$$[S(X; Y; Q) \rightarrowtail \{A^{\diamond}{}_1, A^{\diamond}{}_2, \ldots, A^{\diamond}{}_m, W_{m+1}, W_{m+2}, \ldots, W_n, Q_{n+1}, Q_{n+2},$$

$$\ldots Q_p, D_{p+1}, D_{p+2}, \ldots, D_q\}] \hookrightarrow A^{\heartsuit}.$$

How question. See *Question* and also *Answer*.

Humankind. See *Scale of didactic codeterminacy levels*.

Infrastructure and superstructure. The notion of *infrastructure* (or *substructure*) is, in the ATD, a general concept: it refers to the underlying base needed to develop any determined, superstructural activity. It should be clear, for example, that the "superstructural" activity that consists in watching TV at home requires an enromous infrastructural base. In a school system Σ, the infrastructure allows the appropriate actors of Σ to engage in the superstructural activities of creating and managing the schools σ that the system Σ will consist of. In each of these schools σ

there are also infrastructural means to create and manage classes c, for example by solving problems of time and place of operation. In each class, there are similarly infrastructural devices that allow the superstructural activities that make up the class to be carried out. In a mathematics class, there is a gradually built infrastructure allowing the mathematical (superstructural) activities to be carried out by the students. To be able to write that we have $141217/3215763 = 0.04391... \approx 4.39\%$, for example, we need to have available the division operation and the system $D \geq$ of nonnegative decimal numbers, without forgetting a sufficient calculation time by hand or a calculator, together with the notions of "almost equality" and percentage and their respective symbols ($\approx$ and %). It should be noted that, in many cases, at least within the *paradigm of visiting works*, the time taken to build the mathematical infrastructure leaves relatively little room for the (superstructural) mathematical activities that this infrastructure is supposed to make possible. Things go differently within the *paradigm of questioning the world*, insofar as the mathematical infrastructure is built according to the needs of the superstructural mathematical activities that one wishes to develop. In this perspective, it should be noted that the infrastructure made available by the Internet and digital information technology offers a quite favorable framework to the pedagogies of inquiry.

Inquiry.   See *Dialectics of inquiry* and *Herbartian schema*.

Instance.   In the ATD, the word *instance* refers to either a person x or an institutional position (I, p). In the former case we have a *personal instance* $\hat{i} = x$, in the latter a *positional instance* $\hat{i} = p$. We denote by $R(\hat{i}, o)$ the relation of the instance $\hat{i}$ to the object o.

Instance of reference.   See *Research in didactics and the plurality of instances*.

Institution.   See *Persons, institutions, and institutional positions*.

Institutional break.   See *break*.

Institutional instance.   See *Instance*.

Institutional relation.   See *Cognition*.

Institutional transposition.   See *Praxeologies and praxeological analysis*.

Isodidactic.   See *Didactic gestures and the possibly didactic*.

Lecture.   See *Didactic organisations*.

***Logos* block.**   See *Praxeologies and praxeological analysis*.

Media.   See *Dialectic of conjecture and proof.*

Milieu.   See *Dialectic of conjecture and proof.*

Moment.   A moment is an invariant in the accomplishment of a task t of a given type T; i.e. it is a type of tasks T^* that will necessarily appear in carrying out the task t, whatever the technique (within a certain family of techniques) used to do so. The reason for the choice of the word *moment* is explained by the phrase "there comes a *moment* when": there comes a moment when a task $t^* \in T^*$ has to be carried out, whatever the way of accomplishing task t.

Moment of evaluation.   See *Didactic moments.*

Moment of exploration.   See *Didactic moments.*

Moment of institutionalisation.   See *Didactic moments.*

Moment of the first encounter.   See *Didactic moments.*

Moment of the technique.   See *Didactic moments.*

Moment of the technological–theoretical block.   See *Didactic moments.*

Noosphere.   See *Scale of didactic codeterminacy levels.*

Objects.   See *Cognition.*

Paradigm of questioning the world.   See *Didactic organisations.*

Paradigm of visiting works.   See *Didactic organisations.*

Pedagogy.   See *Scale of didactic codeterminacy levels.*

Person.   See *Persons, institutions, and institutional positions.* See also *Dialectic of persons and institutions.*

Personal instance.   See *Instance.*

Personal relation.   See *Cognition.*

Persons, institutions, and institutional positions.   A *person* is any human being. In particular a newborn, an infant, or a toddler are persons. An *institution* is any created reality of which people can be members (permanent or temporary). For example, a class, a couple, a football club, a bar, a conference, a concert, etc., are institutions. In all institutions, there is a number of *positions*, at least one. In a classroom, there is at least the position of student and the position of teacher; in a bar, the position of client and the position of waiter; in a football team, that of goalkeeper, etc.

Position.   See *Cognition* and *Persons, institutions, and institutional positions*.

Positional instance.   See *Instance*.

Possibly didactic situation.   See *Didactic gestures and the possibly didactic*. See also *Research in didactics and the plurality of instances*.

Praxeological equipment.   See *Praxeologies and praxeological analysis*.

Praxeological universe.   See *Praxeologies and praxeological analysis*.

Praxeologies and praxeological analysis.   A researcher in didactics ξ examines which gestures δ are held, by which instances $\hat{w}$, concerning which cognitive nuclei $\tilde{n}$ under which set of conditions C, to be didactic (or antididactic, or isodidactic), and what changes do these gestures generate in C—including learning and unlearning processes. The verdict pronounced by the instance $\hat{w}$ is based on their relation to ς, i.e., $R(\hat{w}, \varsigma)$, which depends on the relations $R(\hat{w}, o)$, $R(\hat{w}, \hat{\imath})$, $R(\hat{\imath}, o))$, $R(\hat{w}, (I, p))$, $R(\hat{w}, \hat{v})$, $R(\hat{w}, R(\hat{v}, (I, p)))$, $R(\hat{w}, \hat{u})$, $R(\hat{w}, \delta)$, $R(\hat{w}, C)$, etc. Where do these relations come from? What are they born of? What is their genesis? These are the key questions, which can be answered thanks to the notion of *praxeology*. The ATD contains a theory of human action whose starting point is the notion of a *type of tasks* T (and the notion of *task* t of type T, which is denoted by $t \in T$). As is often the case in the ATD, there are no "size" criteria: "scratching your ear", "calculating the difference of two integers", "writing a novel", "reading a newspaper article", "getting married", are all equally types of tasks. The realization of task $t \in T$ requires the use of a *technique* τ_T relating to T. This applies to all actions: walking, singing, eating require learned techniques. In reverse, imitating a person "denaturalizes" his or her behaviour. The ordered pair made up of a type of tasks T and a technique τ relating to T is denoted by $[T/\tau]$ and is called the *praxis block* (or "know-how"). Any technique requires an explanatory or supporting comment, called its *technology*, denoted by the letter θ. The technology θ is itself coupled with a justifying discourse at a higher level, which is the *theory*, denoted by the capital letter Θ, of the technique τ. The ordered pair made up of the technology θ and the theory Θ is denoted by $[\theta/\Theta]$: it is the *logos block* (or "knowledge"). Any *praxis* block $[T/\tau]$ supposes a *logos* block $[\theta/\Theta]$ with which it forms a praxeology $p = [T/\tau/\theta/\Theta]$. Let us set $\Pi = [T/\tau]$ and $\Lambda = [\theta/\Theta]$, we can write: $p = [T/\tau/\theta/\Theta] = [T/\tau] \oplus [\theta/\Theta] = \Pi \oplus \Lambda$. In an institution I, the blocks Π and Λ of a praxeology $p = \Pi \oplus \Lambda$ coming from an institution I_0 are generally modified by the phenomenon of *institutional transposition* of praxeologies (or *works*), which we can write thus: $\Phi: p = \Pi \oplus \Lambda \mapsto p^{\star} = \Pi^{\star} \oplus \Lambda^{\star}$. If δ is a transpositive gesture and $\varsigma = (\tilde{n}, \hat{u}, \delta, C)$

is $\hat{w}$-didactic, we'll speak of $\hat{w}$-*didactic transposition*. All of the above applies to any action. If, thus, $\hat{w}$ judges ς antididactic, this judgment results from a praxeology that allows $\hat{w}$ to make a judgment based on their relations to the many relevant objects o. Where do these relations come from? We define the *praxeological universe* of $\hat{\imath}$ by $\Omega^{\blacklozenge}(\hat{\imath}) \triangleq \{p/R(\hat{\imath}, p) \neq \varnothing\}$ and $\hat{\imath}$'s *praxeological equipment* by $\Gamma^{\blacklozenge}(\hat{\imath}) \triangleq \{(p, R(\hat{\imath}, p))/p \in \Omega^{\blacklozenge}(\hat{\imath})\}$. Note that we have $\Omega^{\blacklozenge}(\hat{\imath}) \subset \Omega(\hat{\imath})$, where $\Omega(\hat{\imath}) = \{o/R(\hat{\imath}, o) \neq \varnothing\}$ is the *cognitive universe* of $\hat{\imath}$, and $\Gamma^{\blacklozenge}(\hat{\imath}) \subset \Gamma(\hat{\imath})$, where $\Gamma(\hat{\imath}) = \{(o, R(\hat{\imath}, o))/o \in \Omega(\hat{\imath})\}$ is the *cognitive equipment* of $\hat{\imath}$. We then posit that, in the reverse direction, $\Gamma^{\blacklozenge}(\hat{\imath})$ generates $\Gamma(\hat{\imath})$ in the following sense: whatever the object o, the relation $R(\hat{\imath}, o)$ results from all the relations $R(\hat{\imath}, p)$, where $p \in \Omega^{\blacklozenge}(\hat{\imath})$, involves the object o, whether technically, technologically or theoretically. The above applies to any object o. Thus the relation $R(x, x')$ of person x to a person x', for example to their own mother ($x' \neq x$), or to themselves ($x' = x$), arises from all the praxeologies to which x has a non-empty relation and which involve x'. The same remark applies to institutional instances $\hat{\imath}$ and $\hat{\imath}'$.

Praxeology.   See *Praxeologies and praxeological analysis.*

***Praxis* block.**   See *Praxeologies and praxeological analysis.*

Question.   Questions are the starting point and the main incentive of didactic life: to ask oneself—or to ask someone—a question is the basic act that will ultimately cause praxeologies as yet unknown to be met. Proper managing of questions is therefore crucial. Two types of questions are mainly considered in ATD. The first type is made up of *how* questions: given a type of tasks T, the *how* question relating to T, Q_T, reads: "How can one accomplish a task t of type T?" An answer to such a question consists essentially in a technique, τ_T, complemented by a technological and theoretical block, $[\theta_T/\Theta_T]$. The second type of questions is that of *why* questions— Why is it so that...? In such a case, the answer consists of an "explanation", which finally refers to some technological and theoretical block, of which it is part and parcel. Note that a *why* question, in fact, conceals a *how* question: "*Why* is it so that...?" is equivalent to "*How* can one explain that...?", that is to a *how* question hinging on the *genre* of tasks "to explain". (Answering such a question, however, will necessitate specifying a *type* of tasks within the genre.) Indeed, most questions, if not all, boil down to *how* questions. For instance, a question about the nature of some entity—What is didactics? What are miscellanies? What is a pupa? Etc.—hides a question about how can one define such entity (the genre of tasks being here "to define"). The same occurs with true-or-false questions—Is it true (or false) that...?— which disguise questions of the "How can one determine the truth value of...?" subtype. Moreover, in ATD, all these types of questions are conditional on the institution to which they are referred: the question "How can one do, explain, define, etc." in fact should be construed as meaning "How in this or that institution do they do, explain, define, etc." (See also *Answers*.)

Reduced Herbartian schema.   See *Herbartian schema.*

Reference instance.   See *Instance*.

Relation.   See *Cognition*.

Research in didactics and the plurality of instances.   The field of research in didactics may be denoted by $\hat{\Delta}$. In $\hat{\Delta}$ regarded as an institution, there exist different researcher positions $\hat{r}$, among which is the position $\hat{p}$ of the researchers in didactics working in the framework of the ATD. Researchers in didactics are generically denoted by the letter ξ. From the point of view of the ATD, the positions $\hat{r}$ (including $\hat{p}$) have neither prerogative nor privilege as concerns their relations to objects o. Whenever $\hat{r}$ is presented as a position of "specialists" of an object o in $\hat{\Delta}$, one may be tempted to think that the "absolute" statement" $R(\hat{r}, o) \neq \varnothing$ actually means: $\hat{r} \vdash R(\hat{r}, o) \neq \varnothing$. The ATD posits that this statement has no special privileges over other statements of the form $\hat{j} \vdash R(\hat{r}, o) \neq \varnothing$. Quite often, the position $\hat{r}$ relies on a position $\hat{p}$ reputed to be that of "true specialists" of o, with $R(\hat{p}, o)$ being the ultimate criterion to judge $R(\hat{r}, o)$. This attitude of $\hat{r}$ towards $\hat{p}$ is a risky one, for it can generate an illusion of mastery (relating to o) in the researchers ξ in position $\hat{r}$, which in turn may stifle ξ's effort to make sense of the reality observed, including $R(\hat{p}, o)$. Now what is of interest to the researcher ξ as such? It is the conditions that can make a relation $R(\hat{r}, o)$ evolve, in particular that can make the instance $\hat{r}$ come to know the object o "better". To move forward, ξ has to take a step aside. To this end, we consider an instance $\hat{w}$ called the *instance of reference*, which can be *any* instance; the prefix $\hat{r} \vdash$ is henceforth replaced by the prefix $\hat{w} \vdash$, and we will therefore ask, for example, whether $\hat{w} \vdash R(\hat{r}, o) \neq \varnothing$ or $\hat{w} \vdash R(\hat{r}, o) = \varnothing$. Of course, we can refocus the analysis on $\hat{r}$ by taking $\hat{w} = \hat{r}$. The introduction of the reference instance $\hat{w}$ opens the way for the introduction of the key notions of *cognitive nucleus* and *possibly didactic situation*.

Scale of didactic codeterminacy levels.   The ATD seems to differ from many other theorisations in didactics in that it does not intend to ignore any of the conditions that possibly exist in a given society. All these conditions are arranged on a scale known as the scale of levels of didactic codeterminacy. The highest level of the scale is that of humanity or human species. The deepest level is that of didactic systems $S(X, Y, \heartsuit)$, where X is the set of *students*, Y is the set of study aids (teacher, etc.), and where the symbol $\heartsuit$ indicates the object which is the didactic stake, "the thing to learn". Too many didacticians are only interested in the conditions that originate at the level of didactic systems proper. Even more restrictively, a teacher often tends to focus on conditions that he or she can create or modify by himself or herself as a teacher. However, we know that what happens in a didactic system cannot be fully explained by the gestures of the $y \in Y$ and the $x \in X$. A didactic system presupposes, first of all, an institution that makes its existence possible, namely a *school* (which can be a family, a sports club, etc.), which itself is included in a society. At this point, the scale has the following structure: *Humankind* $\rightleftarrows$ ... $\rightleftarrows$ *Societies* $\rightleftarrows$ ... $\rightleftarrows$ *Schools* $\rightleftarrows$ ... $\rightleftarrows$ *Didactic systems*. The gestures accomplished, and the conditions created, within a didactic system are called (possibly) didactic *in the strict sense*. The others are (possibly) didactic in the broadest sense. This is the case for "pedagogical" conditions and gestures, that originate at the level of *pedagogies* located between the level of schools and the level of didactic systems. We thus have: ... $\rightleftarrows$ *Schools* $\rightleftarrows$ *Pedagogies* $\rightleftarrows$ *Didactic systems*. To go further, we use the notion of *work* (generically denoted by the letter o). Works are all objects whose existence within a society is due to human action, that is to say... all objects—which are never purely "natural", nor, moreover, created *ex nihilo*. The purpose of a pedagogy is to lead students to the work o to study. For this reason, we see that a pedagogy is by no means independent of $\heartsuit = o$. The existence of a school is thus a pedagogical condition. The grouping of students into classes is also a pedagogical condition—there is nothing "natural" about it. The division of knowledge into disciplines, domains, sectors, themes and subjects of study is another one. These conditions are unequally specific to $\heartsuit = o$, but they all aim to create the situations that will allow the study of $\heartsuit$. The prior arrangement of knowledge in disciplines, etc., is typical of the *pedagogies of visiting works*. It is somewhat different with the *pedagogies of questioning the world*. Between schools and societies, we can then situate the level of the *noospheres*, an institution that, within society, "manages" the organization and future of schools. We then have this: ... $\rightleftarrows$ *Societies* $\rightleftarrows$ *Noospheres* $\rightleftarrows$ *Schools* $\rightleftarrows$ ... Here we include the level of noospheres in the level of societies: ... $\rightleftarrows$ *Societies* $\rightleftarrows$ *Schools* $\rightleftarrows$ ... It then remains to examine a level with a pompous name, the level of *civilizations*: Humankind $\rightleftarrows$ *Civilisations* $\rightleftarrows$ *Societies* $\rightleftarrows$ ... Here, the word "civilization" does not refer to a globality: it has a *local* meaning. So let us assume societies S and S', institutions I in S and I' in S', and positions p and p' in I and I' respectively. (Note that we can have $S' = S$ and even $I' = I$). We will say that, for the instance $\hat{w}$, (I, p) and (I', p') *belong to the same civilisation* as concerns the object o if $\hat{w} \vdash R_I$ $(p, o) \approx R_{I'}(p', o)$. (I, p) and (I', p') *belong to different civilisations* if $\hat{w} \vdash R_I(p, o) \neq R_{I'}(p', o)$. This will be the case, for example, if $\hat{w} \vdash R_I(p, o) \neq \varnothing$ and $\hat{w} \vdash R_{I'}(p', o) = \varnothing$. Given an institutional position (I, p) in a society S, if $\hat{w} \vdash R_I(p, o)$ at time $t_2 \neq R_{I'}(p, o)$ at time t_1, where $t_1 < t_2$, we will say that, between t_1 and t_2, there has been a civilisational change in the eyes of $\hat{w}$ as concerns o—a change *in* civilisation, not a change *of* civilisation. A school, a class are institutions where students are almost constantly confronted with civilisational changes.

School. See *Scale of didactic codeterminacy levels*.

Semi-developed Herbartian schema. See *Herbartian schema*.

Specimen of a type of tasks. See *Types of tasks*.

Student. See *Didactic system*.

Subject. See *Dialectic of persons and institutions*.

Supersession. See *Dialectic of*.

Tasks.   See *Type of tasks.*

Teacher.   See *Didactic system.*

Technique.   See *Praxeologies and praxeological analysis.*

Technological and theoretical block.   See *Praxeologies and praxeological analysis.*

Technology.   See *Praxeologies and praxeological analysis.*

The didactic.   See *Didactic analysis.* See also *Type of tasks.*

Theory.   See *Praxeologies and praxeological analysis.*

Topos.   See *Didactic organisations.*

Transposition.   See *Praxeologies and praxeological analysis.*

Type of tasks.   Every natural language provides an analysis of human activity: when we say "He washes his hands", "She solves the equation for x", "They sing an old lullaby", "The man swims to the buoy", "The mother teaches her three-year-old to blow his nose", we refer to some determined *task* of a certain *type*: to wash one's hands, to solve an equation for x, to sing an old lullaby, to swim to a buoy, to teach one's three-year-old to blow one's nose. A task t of a given type T is called a *specimen* of that type, which is usually written $t \in T$. Grammatically, all these statements are generally made up of a verb of action (to wash, to solve, to sing, to swim, to teach) and a direct object (one's hands, an equation, a lullaby, [the distance] to the buoy, one's three-year-old to blow one's nose). It must be remarked that, in the last case, the "direct object" involves a nested type of tasks, "to blow one's nose"; in fact, the sentence displays the ternary structure typical of *the didactic*: someone (Y, in this case, the mother) does something to help someone (X, her three-year-old) learn something ($\heartsuit$, to blow one's nose). It must also be emphasized that a verb of action (to sing, to walk, to draw, etc.) refers not to a *type* of tasks but to a *genre* of tasks, which will be narrowed into different *types* of tasks by specifying the *object* to which the action applies. Although it is usual to say that one has learnt "to sing", or "to draw", or "to cook", or to "solve equations in one unknown", etc., only *types* of tasks can be objects of learning: *genres of tasks* are beyond reach since any new type of objects can create a new type of tasks necessitating a brand-new technique. Conversely, one cannot learn to accomplish a task regarded as unique: any task has to be recognised as belonging to a certain *type*—i.e. as a *specimen*—so that an appropriate technique can be applied to it (or can be built). One should, therefore, beware not to speak of "the task" when indeed the *type* of tasks is meant.

Why question.   See *Question.*

Work.   A work (French *œuvre*) is any intentional product of human activity. Two types of works are of special interest to ATD. The first one is that of *questions* (which are indeed products of human activity); the second one is that of *praxeologies*. Both categories of works appear in the *reduced Herbartian schema* central to ATD, viz. $S(X; Y; Q) \rightarrowtail A^\heartsuit$, where Q is a question and $A^\heartsuit$ is a praxeology (or a significant part of a praxeology).

ACKNOWLEDGEMENTS

The elaboration of this book has benefited from the support of projects RTI2018-101153-B-C21 and RTI2018-101153-A-C22 from the Spanish Ministry of Science, Innovation and Universities, the Agencia Estatal de Investigación (AEI) and the European Regional Development Funds (MCIU/AEI/FEDER, UE).

PART 1
Unity in Diversity: ATD at Work

1

WHAT KIND OF RESULTS CAN BE RATIONALLY JUSTIFIED IN DIDACTICS?

Josep Gascón and Pedro Nicolás

1 The question of the presumed normativity in didactics

Let us start with a question about the alleged normative character of didactics:

> To what extent, how and under what conditions can (or must) didactics issue *value judgements* and *normative prescriptions* about how to organise and manage study processes?

Most of the approaches and theories in didactics have an implicit position about this question. However, explicit discussions about the legitimacy of didactics to express normative prescriptions remain scarce, let alone specific deliberations about the logical form of the results one should expect. At the same time, discourses in didactics often contain value judgements concerning teaching and learning processes, which sometimes lead to proposals for rules of action—the well-known "implications for teaching". Including value judgements and normative prescriptions in the discourse, often as the main conclusion, appears as a tacit taking of position that has a strong effect on the sort of research questions to be considered and the kind of endorsed answers. It may even lead one to believe that research results in didactics can (or even must) be stated in terms of values and rules.

How can we answer this question according to the ATD's basic assumptions and principles? First, we will explain the ATD perspective about the object of study of didactics and the corresponding research results. This will make clear that value judgements and normative prescriptions have no place in the realm of research results in didactics. Another consequence will be the need to clarify the type of assertions didactics can formulate on solid foundations (Gascón & Nicolás, 2017).

2 Max Weber's thesis on social sciences

The origin of one of the main points of this chapter is (Weber, 1917/2010). In this work Max Weber stands up for the following thesis:

- Social sciences can only state assertions about the rationally suitable *means* to achieve *ends* previously fixed but whose validity cannot be rationally established.
- The precept of avoiding value judgements is a consequence of the distinction between *two spheres of reality*, the knowledge sphere and the values sphere, each one containing questions of different nature.

In the *knowledge sphere*, the main questions raised relate to the behaviour of a certain portion of the world. Why is this portion of the world the way it is? What are the required conditions for this portion of the word to be modified in a certain direction, and which are the obstacles to this modification?

In the *values sphere*, the main questions are about what we should do in a given situation; how we value this situation; if we should do something to change this situation in a given direction, and, if so, in which one.

In the knowledge sphere, it is possible to make a critique of a given value *via* an analysis of the required means to attain this value, regarded as an end. Indeed, one can conclude that, given a school institution *I*, a certain kind of teaching (regarded as valuable, as representing a positive value) is not achievable in *I* because the required means to put it into practice cannot be implemented in *I*. As we can see, this critique cannot state whether this value is estimable or worthy. It can only say that certain means are suitable or not to achieve a certain goal, that certain conditions make it easier or more difficult to attain. Therefore, the object of study of social sciences, according to Weber, concurs with the object of study of didactics according to the ATD sense (see section 4).

The principle according to which science should not make value judgments has a positive reverse: the ability of science to provide criteria to guide us on the steps to follow. These criteria do not permit us to shun the responsibility to make decisions, but they can help us to foresee the consequences of our actions.

Ultimately, the *coercive* aspect of Weber's theses points out the issues and decisions excluded from the scientific scope. Science cannot teach us *what to do*, only *what can be done* and the corresponding consequences. Thus, for instance, economy, as a science, is not qualified to decide how material goods *should* be distributed, in the same way, sociology cannot decide about the *best* social or political structure for societies or physics about what *should* be done with atomic energy. But, in all these cases, and many others, science can reveal some relevant consequences of each possible decision.

3 Reinterpretation of Weber's thesis in epistemological terms

To prepare our description of the object of study of didactics, let us express Weber's theses in terms of explanations and scientific laws. Science is important to society to the extent that it provides laws which support non-trivial and general explanations about the occurrence of objective facts. For a fact to be objective we require it to be *intersubjective*—perceptible by everyone in standard circumstances—and *substantive*—its existence does not depend upon someone's perception or representation.

A *scientific explanation* can be considered as an answer to a question of the type: "Why the occurrence e is the case (instead of the occurrences d_1, d_2, …)?" Of course, deductive arguments are always welcome, but sometimes something different can be accepted as a scientific explanation, for instance, inductive or abductive arguments. The reader is invited to consult (Andersen & Hepburn, 2016; Woodward, 2017) for an overview of this topic.

Typically, a scientific explanation is a valid argument based on a so-called *scientific law*. What a scientific law is, remains a controversial issue in the philosophy of science. We will therefore only present a summary of the most commonly accepted features of scientific laws. The logical structure of a scientific law is as follows: "Every occurrence of type A is also of type B". This logic structure makes it clear that the basic parts of scientific laws are occurrences-type (defined as the extension of a property, or giving an exhaustive list, or by recursion, etc.). Therefore, scientific laws are general statements and not assertions about particular occurrences localised in time-space. Another feature of scientific laws is that they seem to express a *necessary* relationship between types of occurrences, and are not merely *accidental* coincidences.

For instance, the law stating that metals are good conductors should not be regarded as pointing out a fortuitous relationship between the occurrence-type "being metal" and the occurrence-type "being conductive". On the contrary, it seems to say that, given what we know about the world, metals are good conductors and *it could not be otherwise*. This is an important feature of scientific laws, which tells the difference between them and incidental true generalities, as, for example, the statement: "All the mountains on Earth have an altitude of fewer than 9,000 meters." Indeed, the property "being a mountain on Earth" does not seem to imply the property "having an altitude of fewer than 9,000 meters".

As we have seen, it is difficult to find the formal distinction between scientific laws and incidental generalities. Nevertheless, there are other features that distinguish them. Unlike arbitrary general statements, scientific laws are part of a scientific theory in which they are logically linked to other statements. Via its laws and its premises, a scientific theory always describes a certain part of the world. Despite being the compression of the complexity of reality, this description still allows us to master a certain part of the world: answering questions; state predictions; etc.

As an example of a law in didactics, we can consider the statement about the existence of the *didactic contract* (Brousseau, 1997; see also Nicolás, 2015). This law states that, under certain conditions, a study community formed by students and a teacher, along a study process (occurrence of type A), behaves according to specific (perhaps implicit) clauses which rule, among other things, the conduct that students expect from the teacher and vice versa, regarding the knowledge to be taught (occurrence of type B).

For the law of the didactic contract to be scientific, we should show, on the one hand, that both types of occurrences are objective and, on the other hand, that the law itself is objective. To claim this objectivity of the type of occurrences we should provide a sharp, precise description of them. For instance, we should answer the question: what are the precise clauses of this contract? If this is not clear enough, how could we verify the law? To ensure the objectivity of the law we could check its validity in a representative sample of the kind of study communities under consideration, and then use statistical inference. Anyway, we could also consider the law provisionally accepted if we show that, for the moment, it is the best explanation for a certain mysterious or disturbing phenomenon.

The case of the *age of the captain* is an example of such a phenomenon. As explained in Equipe Elémentaire IREM de Grenoble (1979), a research team noticed that students at school tend to operate with the numbers of the formulation of a problem in mathematics regardless their appropriateness for the solution. For instance, when they face the problem "A captain of a ship owns 26 sheep and 10 goats. How old is the captain?", the majority of students give as an answer a number which results from operating with 26 and 10.

In Chevallard (1988) we find an explanation for this phenomenon: students do not pay attention to the meaning of the numbers because their behaviour is governed by a didactic contract including a clause according to which school problems always have a solution, reachable from the numbers appearing in the formulation. Therefore, the law that states the existence of the didactic contract including the clause above appears to be the best explanation for an intriguing phenomenon.

Hopefully, after this brief reflection about scientific laws we are persuaded that the questions placed by Weber in the *knowledge sphere* admit scientific answers (in terms of scientific laws and explanations), unlike the questions placed by Weber in the *values sphere*. Indeed, the occurrences involved in the questions of the values sphere are not objective.

4 The object of study of didactics

The positive part of Weber's thesis is that we have interesting questions falling under the knowledge sphere. Let us formulate these questions to link them to didactics. According to the ATD, the theory of praxeologies is not only rich enough but also especially useful to describe all the human behaviours relevant for didactics. Therefore, in the questions included in the knowledge sphere, we will change the expression "portion of the world" by the term "praxeology" (see Glossary). Moreover, in our formulation, we will take into account one of the basic assumptions in ATD according to which the primary object of study in didactics are the processes of genesis and diffusion of *institutional praxeologies*.

After these considerations, the resulting questions are as follows. Which is the behaviour of a certain *praxeology* in a given *institution*? Why is this *praxeology* the way it is in this *institution*? Which are the required *institutional* conditions for this *praxeology* to be modified in a given direction, and what are the *institutional* obstacles to this modification?

All this is coherent with the *object of study* of didactics according to the ATD: 'the ATD suggests the following definition of didactics: didactics is the science of conditions and constraints for the diffusion of praxeologies in the institutions of the society' (Chevallard, 2011, p. 27, our translation).

To study the aforementioned conditions and constraints for the diffusion of praxeologies and, more explicitly, to account for the *didactic phenomena* linked to that diffusion and which appear in the different institutions of the society, didactics formulates questions belonging to the following three fundamental dimensions of a research problem: *economic, ecological* and *epistemological* (Gascón, 2011).

The *economy* of any system (of the body of an animal, of a plan, of a language, of a discourse, of a book, etc.) is formed by the set of rules and principles which govern the structure and the running of the system. In particular, given an institution, *I*, the *praxeological economy* of *I* at a certain period is given by the set of rules and principles which govern the institutional life of the praxeologies of *I* during this period. Thus, when we tackle the *economic dimension of a didactic problem* in which there are several praxeologies involved, we are concerned with those rules and principles.

The *ecology* of a system is given by the set of conditions and constraints that have an impact on this system. Those conditions and constraints allow an explanation, at least in part, of the evolution of the system, its current behaviour (its economy) and its possible future evolution. As a consequence, when we tackle the *ecological dimension of a problem* in which there are several praxeologies involved, we deal with questions concerning the conditions and constraints which:

- explain the behaviour of the corresponding praxeologies in the considered institution at a certain period (that is to say, the economy of those praxeologies at that moment in that institution),
- are required to promote or impede the life of certain kind of praxeologies in the given institution,
- facilitate or hinder the modification of certain kind of praxeologies in a given direction.

In order to study the economy and the ecology of praxeologies involved in the study of a certain didactic problem, researchers use as a reference both a model of the *praxeologies relative to the knowledge* at stake and a model of the corresponding *didactic praxeologies*—those related to the teaching, learning, study and dissemination of the previous ones. The first model is said to be a *reference epistemological model* (REM, see chapter 6), which accounts for *what* to study, and the second one is said to be a *reference didactic model* (RDM), which accounts for *how* to study. In practice, both models determine each other so we should speak of a *reference epistemological-didactic model* (REDM). The *epistemological dimension of a didactic problem* includes the questions devoted to getting criteria for the construction of a suitable REDM.

The three dimensions of a research problem are mutually determined. Still, in practice there seems to be a certain hierarchy among them. Indeed, to deal with a question included in the ecological dimension—e.g. how to change the conditions for the study of a certain piece of knowledge?—one seems to need certain answers to questions included in the economic dimension—e.g. what is the current state of play with this piece of knowledge?—which, in turn, are based on the (perhaps implicit) assumption of an epistemological point of view—e.g. a certain description of the piece of knowledge at stake (Gascón, 2011, p. 206). Moreover, when we wonder about the current behaviour of certain praxeologies (economic dimension) we have to deal with certain derived questions that can be answered only by researching what happens when we try to change those praxeologies in a given direction (ecological dimension).

We can now reformulate the negative part of Weber's theses in terms of praxeologies and institutions, to get the kind of questions which, being included in the values sphere, cannot be addressed from a scientific point of view. They include questions about how to *value* the presence of certain praxeologies (for instance, describing a very algorithmic and rote-learning approach to teaching mathematics) in a given institution; about changing the presence of these praxeologies in this institution in a given direction; or about deciding which one of two didactic praxeologies seems to be the best.

5 Nature of didactic results in the ATD

In the ATD, research results—that is, the (tentative) answers to research questions—can be formulated in terms of didactic laws describing certain didactic phenomena. The ATD provides different *didactic laws* whose formulation relies on the reference epistemological-didactic model used in each case. Didactic laws addressing the *economic* dimension relate to how certain praxeologies behave in a given institution. Let us give two examples of statements that are good candidates to become didactic laws.

Example 1a: In secondary education in Spain—and, most likely, in similar school institutions—mathematical praxeologies about *proportionality* have given rise to a general pervasive epistemological-didactic model that isolates mathematical praxeologies about proportionality from other mathematical praxeologies concerning functional relations (occurrence of type A). In the corresponding didactic praxeologies one finds certain phenomena like the *avoidance of algebra* and the characterisation of proportionality as a *purely arithmetical* relationship (occurrence of type B) (García, 2005).

Example 2a: The mathematical praxeologies about *elementary differential calculus* in secondary education in Spain—and, most likely, in similar school institutions—have given rise to a general pervasive epistemological-didactic model in which praxeologies are separated from the activity of *functional modelling* (occurrence of type A), which consists of using elementary functions as models to provide answers to certain questions. Consequently, the corresponding didactic praxeologies do not attribute to elementary differential calculus a raison d'être related to the construction, manipulation and interpretation of different kinds of functional models (occurrence of type B) (Lucas, 2015).

Didactic laws addressing the *ecological* dimension give an account of the conditions that made possible the current state of the praxeologies (the economy), the conditions required to change those praxeologies in a certain direction and the constraints that hinder this change. To state those laws, ATD uses, among other things, the *theory of didactic transposition* (see Glossary), which enlarges the field of study and provides insights about the constraints which affected, and still affect, the *taught knowledge* (Chevallard, 1985/1991). Another important tool is the *scale of levels of didactic codeterminacy* (Chevallard, 2011, see Glossary), which enlarges the kind of conditions and constraints researchers in didactics must consider.

An epistemological-didactic reference model is a conjecture (a scientific hypothesis) that provides a (partial) answer to the questions appearing in the three dimensions. In the epistemological dimension, this model *redefines* the piece of knowledge at stake (in particular, it might attribute to it an *alternative raison d'être*). In the economic dimension it serves as a reference to compare with and characterise the *pervasive* epistemological-didactic model. Finally, in the ecological dimension it *conjectures* that certain conditions will permit to avoid some phenomena appearing in the mathematical-didactic organisation of the knowledge at stake.

In this sense, reference epistemological-didactic models are tools by which the researcher emancipates her/himself from the prevailing epistemological and didactic models (Gascón, 2014). Moreover, they also state implicitly a didactic *law*: if in a given institution one follows a study process based on this reference epistemological-didactic model (occurrence of type *A*), then along this study processes certain didactic phenomena will (not) take place (occurrence of type *B*).

Let us present, in what follows, examples of possible *didactic laws* corresponding to the ecological dimension.

Example 1b: If in the final courses of secondary education in Spain (14–16 years) one carries out a study process about *proportionality* based on the epistemological-didactic reference model presented in García (2005), in which the study of proportionality is integrated into a general study of the possible functional relations between two magnitudes (occurrence of type A), then the didactic phenomena described in the example (1a) before will not take place (occurrence of type B).

Example 2b: If in the transition from secondary to tertiary education in Portugal, France or Spain one carries out a study process about *elementary differential calculus* based on reference epistemological-didactic model presented in Lucas (2015) (occurrence of type A), then the didactic phenomena described in the example (2a) before will not take place (occurrence of type B).

Why might one be interested in the elimination of certain didactic phenomena as the ones presented in the two previous examples? Inevitably, as any other didactic approach does, the ATD assumes certain principles concerning which are the *ends* of education. This assumption already conditions the way ATD detects and values phenomena. Of course, the fact that certain ends of education are prioritised among others belongs to the values sphere, so cannot be rationally founded.

6 Enquiring into the relationship between didactic research and teaching

As laws in sociology describe some features of the social world and laws in economy describe some features of the economic world, laws in didactics describe the behaviour of institutional praxeologies. Those laws are general statements that can be used to sustain a good explanation for certain (social, economic, didactic) phenomena, but they never claim value judgments or normative prescriptions or proscriptions.

However, some expressions appearing in works by people using the ATD—for instance, "to turn the study of proportionality into something *meaningful*", "*true* raison d'être of the elementary algebra", "*undesirable* consequences of the disintegration of school mathematics"—can be disturbing and misleading, as they assume value judgments which, in turn, induce normative prescriptions or proscriptions concerning teaching.

Actually, these judgments and norms, although frequent in ATD work, cannot be presented as research results, not even as consequences of them. Research results belong to the knowledge sphere and teaching proposals belong to the values sphere, to the extent that "proposing" entails a taking of positions concerning the ends of education. Thus, when a theory in didactics fosters certain teaching proposals, it is necessarily assuming certain ends.

The assumed ends of education are part of the principles embraced by a theory in didactics. By "principles" we mean those unquestioned statements on which the whole theory rests. Among these principles, we find not only the aims of education but also an ontological description of the part of the world the theory deals with together with methodological premises.

To raise those principles may help to not only illuminate the link between research and teaching proposals but also increase the degree of self-awareness of the theory. Only in this way, with our principles explicit, is it possible to distinguish between research results and values and make clear to what extent we help to enlarge our knowledge of the didactic dimension of human societies.

References

Andersen, H., and Hepburn, B. (2016). Scientific method. In E. N. Zalta (ed.), *The Stanford encyclopedia of philosophy*. Stanford, CA: The Metaphysics Research Lab, Center for the Study of Language and Information, Stanford University.

Brousseau, G. (1997). *Theory of didactical situations in mathematics: Didactique des mathématiques 1970–1990*. Dordrecht, The Netherlands: Kluwer.

Chevallard, Y. (1985/1991). *La transposition didactique: Du savoir savant au savoir enseigné*. Grenoble, France: La Pensée Sauvage (2nd edition 1991).

Chevallard, Y. (1988). *Sur l'analyse didactique: Deux études sur les notions de contrat et de situation*. Marseille, France: IREM d'Aix-Marseille.

Chevallard, Y. (2011). Quel programme pour l'avenir de la recherche en TAD? In M. Bosch *et al.* (eds), *Un panorama de la TAD* (pp. 33–55). Barcelona, Spain: Centre de Recerca Matemàtica.

Equipe Elémentaire IREM de Grenoble (1979). Quel est l'âge du capitaine? *Grand N*, 19(4), 63–70.

García, F. J. (2005). La modelización como herramienta de articulación de la matemática escolar: De la proporcionalidad a las relaciones funcionales (Doctoral dissertation). Universidad de Jaén, Spain.

Gascón, J. (2011). Las tres dimensiones fundamentales de un problema didáctico: El caso del álgebra elemental. *Revista Latinoamericana de Investigación en Matemática Educativa*, 14(2), 203–231.

Gascón, J. (2014). Los modelos epistemológicos de referencia como instrumentos de emancipación de la didáctica y la historia de las matemáticas. *Educación Matemática*, 25, 99–123.

Gascón, J., and Nicolás, P. (2017) Can didactics say how to teach? *For the Learning of Mathematics*, 37(3), 9–13.

Lucas, C. (2015). Una posible razón de ser del cálculo diferencial elemental en el ámbito de la modelización functional (Doctoral dissertation). Universidad de Vigo, Spain.

Nicolás, P. (2015). Structuralism and theories in mathematics education. In K. Krainer & N. Vondrová (eds), *Proceedings of the Ninth Congress of the European Society for Research in Mathematics Education*. Prague: Charles University, Faculty of Education.

Weber, M. (1917/2010). *Por qué no se deben hacer juicios de valor en la sociología y en la economía*. Madrid: Alianza Editorial.

Woodward, J. (2017). Scientific explanation. In E. N. Zalta (ed.), *The Stanford encyclopedia of philosophy*. Stanford, CA: The Metaphysics Research Lab, Center for the Study of Language and Information, Stanford University.

2

RESEARCH ON ATD OUTSIDE MATHEMATICS

Caroline Ladage, Marianne Achiam and Martha Marandino

1 Introduction

Since its first developments in the 1980s, the theory of *didactic transposition* (see Glossary) elaborated in the field of mathematics education has met with success among didactics researchers and education professionals spanning multiple school-related content areas—such as languages, science, history, geography, sports education—as well as various educational contexts outside school, including vocational training. Among the multiple questions prompted by this theory, two have turned out to be especially significant for both researchers and educational professionals: the question of identifying the bodies of knowledge regarded as worthy of being taught in a specific societal context and institution; and the question of the nature of the transformations undergone by knowledge in order for it to be taught. Research has shown that for any given subject matter the question of the precise nature of the knowledge taught (the "didactic stake", the thing to be taught and learned in a given didactic system) is rarely questioned, even more rarely problematised, and rather taken for granted. This is a major issue to be studied in educational sciences, as we will see in the second part of this chapter. First, we want to highlight the widespread denial of transposition processes, be they "didactic" or otherwise, with the case of science museum exhibits.

Science museums have educational practices that give rise to diverse and idiosyncratic products, e.g. exhibitions and programmes (Achiam & Marandino, 2014). To study the inherent mechanisms at work in the development of museum exhibitions, the theory of didactic transposition has proved a useful framework. For example, Mortensen (2010) reports on research on the process of didactic transposition in the development of a museum exhibit on animal adaptations to darkness. The framework served to conceptualise the museum-specific transposition, designated as museographic transposition, with a three-stage framework including the contexts of the scientific source knowledge, the curatorial brief and the exhibition milieu. The method allowed for the mapping of the changes between these stages and the systematic tracking of the epistemological and semiotic changes in the given body of knowledge in the exhibit development process. In this way, the didactic transposition theory more generally allowed for the analysis of the knowledge transformation in exhibition engineering processes and became an analytical method applicable to the development of new exhibits as well as the post hoc analysis of existing exhibits. Further research (Marandino & Mortensen, 2011) reviewing the literature showed that the didactic transposition framework had gradually developed in three directions, with a focus on epistemology (where does the knowledge come from?), semiotics (what can be said about the effectiveness and mechanisms of the produced milieu in promoting learning?) and sociology (how and by whom the knowledge is shaped in its trajectory towards the exhibition?).

As the anthropological theory of the didactic (ATD) developed over the years, it gradually came to include the analysis of the conditions that promote or hinder the diffusion of knowledge in a broad sense. This new perspective considered these conditions on a scale of levels of didactic codeterminacy (see Glossary); the ATD framework thus enriched was also of interest for museum didactics. The use of this scale proved to be particularly interesting as a means of carrying out cross-cultural comparisons of museum practices in a systematic way. This was the case with the comparison of representations of science in museum exhibitions in Brazil and Denmark, which we will develop below. The use of the scale of levels of didactic codeterminacy illustrates how conditions and constraints (see Glossary) originate and manifest themselves at different levels in the two contexts, suggesting that institutional idiosyncrasies originate not just within institutions but also externally.

2 ATD and science museum didactics

At first glance there seems to be a general, international consensus regarding what a museum is. Natural history museums all over the world disseminate roughly similar content in roughly similar ways; the same can be said about science centres or zoos or aquaria—institutions that we will collectively designate as "science museums". However, research shows that the work of creating dissemination activities in science museums is strongly influenced by the cultural and societal contexts in which the work takes place. Individual societies and communities have patterned ways of seeing, valuing, ascribing meaning to and treating objects (Kreps, 2003), and it follows that the work of science museums to inspire educational and communicational processes are the highly diverse and idiosyncratic products of not only their individual, institutional cultures but also the influences of the societies that surround them (Achiam & Marandino, 2014). It is the work of *comparative museology* to identify the similarities and differences in museological practices across cultures (Kreps, 2006); here, we suggest Chevallard's scale of levels of didactic codeterminacy as a means to carry out such cross-cultural comparisons in a systematic and principled way. Accordingly, the purpose of this study is to provide a first example of how the scale of levels of didactic codeterminacy can be used to compare science museum practices across cultures and to argue the merits of this approach.

In this study we compare the conditions and constraints that codetermine the representation of science in museum exhibitions in two different countries, Brazil and Denmark, based on the analysis of two exhibits with similar biological content from two museums of zoology, one in the University of São Paulo (USP) in Brazil and one in the University of Copenhagen (UCPH) in Denmark. Although the ideological project of comparative museology is to "liberate" the interpretation, representation, and dissemination of content from the management of Eurocentric museology (Kreps, 2006), we have chosen to compare two museums from the Western world. Our project is thus not an ideological one; rather, we decided to compare two museums with similar Western backgrounds because the benefits of establishing a first working model of the comparative framework by using well-known (to us) museum cases outweighed the disadvantages of confining ourselves to Western culture and thereby precluding ourselves from recognising non-Western museological behaviour.

In the case of science museums, the scale is adapted to the context of museum exhibits (Achiam & Marandino, 2014), with the following levels: civilisation, society, museum, pedagogy, discipline, exhibition cluster, exhibit, task, and visitor knowledge. It may be useful for the reader to recall that, just as in the general version of the scale, the museum-based version is based on didactic practice as it is realised in the specific institution in question. This means that instead of the levels of subject, theme, sector and domain familiar to ATD adherents, the museum-based scale operates with the levels of task, exhibit, cluster, and exhibition because this is the way museums commonly organise content.

2.1 The case of two zoology museums

The Zoology Museum of the University of São Paulo (ZM-USP) has one of the largest zoological collections in Latin America and its role is to contribute to the development of knowledge about the Brazilian and global biodiversity. With over ten million preserved specimens, the museum gives a unique testimony of species and ecosystems, some of which are extinct today. As stated on its website, ZM-USP sought to follow the general museum tendency of the late twentieth century, emphasising communication with the public and developing the long-term exhibition "Research in Zoology: biodiversity under the eyes of the zoologist". This exhibition was intended to offer visitors the opportunity to experience aspects of the work of zoologists. The exhibition emphasises the importance of the research of zoologists, the natural processes that promote morphological, genetic and ecological diversification of animals throughout the history of the planet, as well as the standards that led to the current distribution of animal species between different environments and continents. The exhibition promotes an institutional narrative, which entails presenting the taxidermied animals in a contextualised way. The inclusion of dioramas of certain Brazilian ecosystems was an attempt to represent the existing organisms in these environments and the ecological relationships between them.

The Zoology Museum of the University of Copenhagen (ZM-UCPH) has a collection of about ten million specimens, representing about 10 per cent of the world's known multicellular species. These collections are pivotal for the Museum's own research efforts but are also much in demand for research conducted at other institutions, both in Denmark and abroad. The exhibition "Danish Fauna: From mammoth steppe to cultural steppe" represents a journey through time, from 20,000 years ago when the glaciers of the latest ice age were pulling back to the present-day cultural landscape of Denmark. The exhibition features the remains of several extinct prehistoric animals and a number of extant animals. A large number of the exhibits in the exhibition are dioramas, in which taxidermied animals are shown in their ecological and historical context.

2.2 Method

To determine what conditions and constraints influenced the production of the exhibitions at the two museums, a range of different kinds of data were collected. Semi-structured interviews were carried out with the two curators at each museum who were responsible for conceiving the exhibitions. All the available documentation of the exhibitions, such as planning briefs, exhibition catalogues, and pamphlets were collected, and the exhibitions were photographed in detail. The collected data were analysed using the *theoretical thematic analysis* procedure described by Virginia Braun and Victoria Clarke (2006). A "top-down" approach was used, in which the levels of didactic codeterminacy described above comprised our deductive framework. Briefly, this approach entailed familiarising ourselves with the data, coding the data using the ten levels of codeterminacy as our focus, allocating the codes to themes, and reviewing the themes to ensure internal consistency.

2.3 Similar and distinct conditions and constraints

In both cases we found clear evidence of conditions and constraints at all levels of didactic codeterminacy. However, even though the development of exhibitions in both cases was conditioned and constrained at all levels, these conditions and constraints did not always *originate* or *manifest* themselves in the same way in the two contexts. Thus, there is evidence that exhibitions, as cultural and social productions, are indeed influenced by their local contexts (Marandino, 2016). In the following, we give the most significant examples of our findings.

2.3.1 Civilisation

The institution of the museum is a product of Western culture (Hooper-Greenhill, 1999). It is deeply enmeshed in this culture, to the extent that the conditions and constraints posed by culture become all but invisible to the actors within the museum. This was evident when the Danish staff member discussed the practices of ZM-UCPH:

> In a little country like Denmark, we would be looking at greater museums like the Natural History Museum in London, the Natural History Museum in New York, the Smithsonian, and Paris, all that. And they would do it in such a way, so I think that's pretty much how to do it.
>
> *(Staff member, ZM-UCPH)*

This quote seems to indicate that there is a kind of consensus about what a natural history museum is; a consensus that spans a number of countries in the Western hemisphere. The staff member from ZM-USP may be referring to this consensus as well when he describes his institution in the following way: "the Zoology Museum is a museum … of natural history, it comes from this lineage". Arguably, this consensus constrains and conditions the practices of those museums that subscribe to it; yet these constraints and conditions may be taken for granted by museum actors and remain more or less tacit (cf. Achiam & Marandino, 2014).

2.3.2 Society

In recent years, museums have become increasingly dependent on securing external financial support as public subsidies decrease (McPherson, 2006). It is therefore not surprising that we found conditions related to financial support to be limiting for the activities of both museums, even if the specific economic contexts of Brazil and Denmark are quite distinct from one another. It seems that today, museums face important financial challenges to justify their existence in any society (cf. Black, 2012).

Another way in which conditions at the level of society influenced the practices of the two museums was evident in the way they established their relationship with their local communities. The museums' expectations of their audiences on the one hand and the interests and requirements of the local visitors on the other revealed that the roles of the museums are linked to the social and cultural aspects of Brazilian and Danish society, respectively. The staff member from ZM-USP explained:

> So, the museum had this goal, a more functionalist perspective, to give answers to these questions that were important to the people who lived here [São Paulo State]. What are the situations that we had to fight in the agricultural, medical, and veterinary [areas]?
>
> *(Staff member, ZM-USP)*

This attention to providing answers to socio-scientific issues is not as evident in the discourse of the staff member from ZM-UCPH, who seems to invoke the museum's obligations as a state-funded institution to disseminate zoology:

> You have an exhibition, you want to show a story about the natural history of Denmark through time, and this being a zoological museum, you want to show the animals.
>
> *(Staff member, ZM-UCPH)*

2.3.3 Museum

At this level, we expected to find conditions and constraints about the fact that both institutions are university museums and thus have dual research and teaching purposes. However, this affiliation seemed to have a much more discernible impact on the practices of ZM-USP, which, when it came under the direction of the university, had to align its role to the institutional "tripod" of Brazilian universities: research, teaching and outreach (institutional document, ZM-USP). Among other things, this condition manifested itself in the choice of the theme of the exhibition:

> So the current exhibition of the Museum of Zoology is dedicated to research in zoology, right? [...] How is research in zoology carried out? Research in zoology can only be carried out in a museum environment, [because] you need the collections. [...] The idea here is to show how research in zoology is done with the right assets, with objects, with natural objects.
>
> *(Staff member, ZM-USP)*

In Denmark, the institutional affiliation of ZM-UCPH as a university museum did not manifest itself as strongly. To be sure, the role of ZM-UCPH as an institution with both research and education functions is clearly stated on the institution's web page, but unlike ZM-USP there is no attempt to communicate the interdependence of the two functions through the exhibitions.

2.3.4 Discipline

In both museums, we found evidence of conditions and constraints originating at the level of discipline. No doubt, this is a direct consequence of the general role of the museum as an educational institution (cf. ICOM, 2007) and the specific role of natural history museums as disseminators of biology-related content. In the case of ZM-USP, biodiversity and its associated notions had a prominent role in defining the content of the exhibits:

> In conclusion, then, there was a contribution [that is] the incorporation of ecological and genetic theories to understand the phenomena that generated the tropical biodiversity, particularly neotropical, which is where we are, right? So we decided to express it using the dioramas.
>
> *(Staff member, ZM-USP)*

The staff member from ZM-UCPH also emphasised how disciplinary knowledge (ecology, in this case) was a strong condition to produce a series of dioramas showing different habitats:

> But in those days, [...] it was ecology that was modern in biology. We had to tell children, people, that every animal lives in a different place, they eat each

other, and all depend [on each other] ... . So animals have demands to their
landscape, and ... that was modern in those days.

(Staff member, ZM-UCPH)

In both cases we can see how conditions originating at the level of the biological
discipline manifest themselves in specific decisions being made at the level of the
exhibit, namely, the choice of dioramas.

2.3.5 Exhibition

As their titles indicate, the two exhibitions represent two different (but over-
lapping) bodies of biological knowledge. Part of the Brazilian exhibition "Research
in zoology: Biodiversity under the eyes of the zoologist" consists of a series of
dioramas that reflect the biodiversity of a number of South American biomes, e.g.
the Amazon rain forest, the Atlantic forest, the *Cerrado* (savannah), etc. This
geographical diversity became directly mapped on to the available exhibition space:

> I worked a lot with planning [...] I applied the map of South America, you
> know, this is the neotropical fauna on display, and I saw that positioning the
> dioramas it would be kind of more or less the right map region, then the Cerrado
> would be very close to where it is usually the Cerrado area [on the map], the
> Amazon where it is, the Amazon.
>
> *(Staff member, ZM-USP)*

In contrast, part of the Danish exhibition "Danish fauna: From mammoth steppe to
cultural steppe" consists of a series of dioramas that illustrate 'time travel through
20,000 years—from the steppe of the mammoth to the present-day cultural landscape' (web page, ZM-UCPH). Thus, the exhibition is laid out as a convoluted,
unidirectional passageway among dioramas that illustrate the succession of animals,
plants, landscapes and climate in Denmark in the last 20,000 years.

2.3.6 Exhibit

We discussed, in the section on Discipline, how the intention to disseminate knowledge
related to ecology and biodiversity manifested itself at the level of the chosen exhibit
genre (the diorama) of both exhibitions. This is not surprising given that, historically,
dioramas were developed specifically to represent ecology-related concepts by
embodying the natural interactions between plants, animals and climate (Marandino,
Achiam & Oliveira, 2015; Van Praët, 1989). In other words, the diorama genre has
strong didactic specificity concerning the mediation of ecology-related knowledge. Staff
members from both institutions agree on the modality of the diorama:

> The dioramas [...] reproduce the natural environment; the taxidermy that is
> made is artistic taxidermy. The animal is placed as if alive, then, with glass

eyes, with open wings, sometimes with raised leg, playing positions that they had while alive.

(Staff member, ZM-USP)

To make a very natural-looking diorama, so it would be like looking into the nature and seeing exactly the same as in the nature. If you want to show nature, you have to copy it. The intention is to make as exact a copy of nature as possible.

(Staff member, ZM-UCPH)

2.3.7 Task

Conditions and constraints at the task level are those that pertain to the actions suggested to the visitors by the design of the individual exhibits. Through our interviews with staff members, it became clear that ZM–USP intended visitors to comprehend certain biological concept—namely, the relation between different organisms and between the organisms and the environment:

The diorama is a scenario where you juxtapose elements, and you want the public to understand the relationships between these elements.

(Staff member, ZM-USP)

The staff member from USP referred to the chosen exhibit genre—the diorama—as the main constraint regarding what could be disseminated; this constraint manifests itself in the types of interrelationship that could be illustrated in the diorama. In contrast, ZM–UCPH seemed to be less ambitious concerning its intentions for visitors. When discussing the specific content of one of the dioramas, the staff member explained:

It should be a typical Danish location, and well, it's a nice picture to show people. So, I don't think there's any other explanation. When … there's a seashore with gulls and terns—a lot of them, that should say: "Wow, now I'm on the seashore"—that's the only thing.

In this case, the positioning of the elements of the diorama was conceivably constrained by the intent of creating a recognisable reproduction of a typical Danish landscape.

2.4 Discussion

Taken together, our results are evidence of the various conditions and constraints that co-determine the production of exhibitions in museums. We showed how these conditions and constraints originated at various levels of didactic codeterminacy and manifested themselves in the production of the exhibitions in both museums studied here. In both cases, several conditions and constraints originated *outside* the museum; this offers support to the finding that "the constituting dynamics of the expositive discourse is also the result of broader political and social movements, which in turn also have a historical dimension" (Marandino, 2016, p. 32). As pointed out by Christina Kreps (2006, p. 312):

> Museums and museological work do not exist in a vacuum but are part of larger sociocultural systems that influence how and why curatorial work is carried out. Because curating cannot be divorced from these contexts, it seems appropriate that scholars and museum practitioners are redefining curating in terms that acknowledge the social and cultural dimensions both of objects and curatorial work.

In other words, even though the interviewed staff members recognised their respective institution as belonging to the same category, namely that of natural history museums, this "membership" does not make them immune to the effects of their surrounding culture and society. Thus, even though the institution of a natural history museum may be Eurocentric (cf. Kreps, 2006), the national and institutional idiosyncrasies detected in this study may be evidence of the partial "liberation" of the natural history museum from the constraints of its origins and its increasing degree of adaptation to more local conditions.

In addition to the external conditions and constraints, the present study also detected several constraints and conditions originating internally, within the walls of each museum. This is not surprising in itself; however, in the comparison between the two museums, these differences suggest that there is substantial idiosyncrasy at work in the internal workings of museums. Specifically, our results seem to point out that the practices of ZM-USP seem to be more strongly conditioned by a curatorial intention of directly addressing the local audience and providing them with answers to questions about biological and zoological research, while the internal practices of ZM-UCPH seem to be more oriented towards the institution's societal obligations as stated at the policy level. However, a more detailed analysis of the available data would be necessary to investigate this hypothesis.

The second outcome of this study is that it provides evidence that the ATD, and specifically the scale of levels of didactic codeterminacy, can be a useful theoretical tool to develop and carry out comparative analyses in the field of museology. Much of the work being done in museum education research tends to focus on phenomena that originate or manifest themselves at either the higher levels of the scale or the lower levels; we agree with Artigue and Winsløw (2010) that without due consideration of conditions and constraints at multiple levels of codeterminacy, such studies risk drawing simplistic conclusions. We therefore suggest that using the scale of levels of didactic codeterminacy could open a whole line of work in comparative museology that could consider each level. Moreover, thanks to the ATD notion of praxeology, the idea of praxis and logos can contribute to research that compares different practices, institutions and nations. It is also within this framework that the relationship between the design of a museum exhibit and the subsequent visitor interactions with and understandings of that exhibit can be studied, using the stated learning objectives for the exhibit as a measure of how well the exhibit performs (Mortensen, 2011).

The study of the museum didactics case within the ATD framework in its earliest as well as latest development gives examples of how mechanisms can be revealed in any type of institution; mechanisms that determine the diffusion of knowledge within a given society. At this point of our analysis on ATD research outside mathematics we recall the general definition of Chevallard, which considers that the aim of didactics as a science is to elucidate the mechanisms by which, in a given society, knowledge is diffused within institutions and among persons—i.e. essentially the diffusion of praxeologies. Studying didactics as a researcher, a pre-service teacher, a teacher, an educator or indeed any professional related to education calls for in-depth studies of praxeologies in given social situations. It is therefore important to learn how to analyse any social situation to be able to reveal, understand and organise didactic mechanisms. This is the issue addressed by the second part of this chapter.

3 In search of didactic material on the internet: a training [and research] tool for didactics at university

The second domain in which the didactic framework is studied in this chapter is educational sciences, where we consider didactics as not only a learning and training object for students but also as a research tool. To prove this threefold utility of the ATD we study a case of didactical engineering that uses didactic analysis as a training tool for students within the context of a *basic didactic* course at university in educational sciences. First, we must stress that in this context, didactics is learnt as a theory, the aim of which is to understand didactic phenomena in a broad sense, disregarding disciplinary specificities. As such we regard the students as future didacticians because they will, in the future, work either as teachers or as researchers in the field of education.

As with most subject matters, theoretical lessons are associated with practical exercises, which, in the case of ATD for educational sciences, are mostly didactic analyses as well as praxeological analyses, plus, more occasionally, course work on didactic transposition and engineering. In the case we study here, didactic analysis, closely associated with praxeological analysis, is regarded as the main didactic stake of the course. Both are central to the ATD as tools designed to come as close as possible to the didactic in any social situation. Yves Chevallard (2011) suggests that the capacity to engage in and carry out a didactic analysis has much wider interest than is generally thought: a "public" interest. Therefore, working on didactic analysis with a group of students does not just concern an analysis of a given didactic situation within a school context. The aim of the work is to put them in contact with the didactic elements all around them and to offer them theoretical tools for understanding the mechanisms involved. The point of such teaching is therefore not to teach didactic analysis as such, with its set of stan-dardised technical procedures applied to chosen disciplinary didactic situations, which one would only be authorised to perform if one was a specialist or at least a student in the field in question. The stake, above all, is to reveal the question of how to find oral or written descriptions of complete didactic situations, high-lighting more or less explicitly the work of people following one or several didactic systems based on didactic questions or stakes, which are also more or less difficult to identify.

As we will see in this study, this identification is not always straightforward once you leave the surroundings of educational institutions. Very often, what can easily be found are documents (of any kind) that offer a description of what we call incomplete didactic situations, showing only limited information on didactic facts.

The research approach thus implemented experimentally finds its full theoretical justification in the concepts well known in ATD central to the paradigms of "visiting works" and "questioning the world". In what follows we explain how we carried out an experiment with students in educational sciences whose task was to complete a didactic analysis of a didactic situation they had to find by themselves. The story of their search and the didactic analysis they produced are analysed in the light of the study of the conditions and constraints affecting the paradigm of questioning the world, highlighting the need to learn to identify, to see, to spot and ultimately to fully analyse the didactic in order to improve its social efficiency. We propose initially to broaden the theoretical approach in which this work of searching for and questioning of didactic situations is enshrined.

3.1 The required ATD tools

How can we analyse the didactic elements present in a situation, either through direct observation or through an oral or written description of a more or less accurately described situation? In order to understand the complex procedures of didactic analysis as developed and taught by Yves Chevallard, we rely on a collection of lessons which he compiled for his education sciences degree and master's students at Aix-Marseille University: the *Leçons de didactique* [Lessons in didactics] from the course *Didactique fondamentale* [Basic didactics] for the year 2011/12 (Chevallard, 2012), and the different notes for the students to help them carry out their didactic analyses, as well as the journal for the ATD seminar (Chevallard, 2011). We begin by recalling the definition given to the students of the *didactic situation* concept (based on Chevallard, 2012, p. 4, own translation):

> You would say that, in a given social situation, there are didactic elements when someone or, more generally speaking, some instance (a person or an institution) envisages doing (or does) something for someone or some instance to learn something. All the words are important in this formulation. The definition proposed is in fact very broad. When a social situation contains didactic elements (or what can be called "didactic facts"), you would say to be brief that it is a didactic situation.

Let us also recall that the structure of the didactic situation (its didactic model) is formalised in the concept of didactic system, a system noted as $S(y, x, \heartsuit)$, where y is a person or institution that does something with the intention of helping x, a person or institution, to study a work, the didactic stake, denoted by the symbol of a heart $\heartsuit$, to express that it is central and most dear to the aim of study. In short, both y and x are also named "instances" and can involve more than one person on both sides. In that case, we use capital letters (Y, X). In its broad scope, didactic analysis will begin with the identification, in the social situations examined, of didactic systems, social realities which include these three entities: a "pupil" x, a "teacher" y, and a didactic stake ♥. The words "pupil" and "teacher" are placed between quotation marks to highlight the fact that the didactic system $S(Y, X, ♥)$ applies well beyond the ordinary academic institution.

The first moment of the didactic analysis work with students is therefore devoted to learning how to find didactic elements around oneself and to analyse them in social situations where they are generally hidden to most observers or immersed beyond clear view. This task is not easy in a society marked, as Chevallard says (2012), by a cultural repression of the didactic: "It is likely that the traditional and unfavourable bias against the adjective didactic is linked to the cultural repression of the didactic: both appear as characteristics of civilisation which are correlated, enduring and ubiquitous" (p. 12, own translation). This repression of the didactic is expressed by the fact that the vast majority of oral and written discourses and accounts on the social world pay no attention to the didactic.

Once the basic schema of a didactic system observable in its entirety has been identified in a social situation, the didactic analysis is based on two theoretical tools central to ATD: the scale of levels of didactic codeterminacy (for studying the conditions and constraints affecting the formation and conduct of the didactic situation), and the Herbartian schema (see Glossary):

$$[S(Y, X, Q) \hookrightarrow \{ A_1^◊, A_2^◊, ..., A_m^◊, W_{m+1}, W_{m+2}, ..., W_n \}] \rightarrow A^♥.$$

The task the students must carry out is guided by an open series of questions: what didactic elements are present in the situation? Which are the didactic systems present or referred to by the people taking part in the situation? What didactic actions are accomplished or envisaged? Which is the mandating institution? What types of tasks, techniques, technologies and theories are incorporated in the situation? What can be said of what the participants learn from the construction of the answer $A^♥$?

Let us note that the analysis of the didactic stake (♥) of a didactic system $S(Y, X, ♥)$ calls for further examination. This is where the didactic analysis meets the need for a praxeological analysis, with the aim of helping to re-problematise the praxeologies referred to in the didactic stake. The Herbartian schema that introduces a question Q at the place of the symbol of the heart (♥) thus introduces a complete change in the functioning of the didactic system necessary to construct an answer $A^♥$. As Chevallard often notes, working on didactics requires one to be capable of "dissecting", of "reading" in the flows of human activity—whether they be mathematical, culinary or something else—combinations that are sometimes incongruous, often complex, with innumerable types of tasks, and to look for clues and signs of the bringing into play the determined techniques.

The praxeological analysis is very evidently essential for didactic analysis. However, it is a difficult and unusual task for students. In what follows we study didactic and praxeological analyses as practical exercises in the already described basic didactic education course as a first step in using ATD as a tool. We will also use praxeological analysis as a tool of the didactic researcher to consider the quality of the results of the assignments that were carried out.

3.2 Quests for didactic situations

We studied the work carried out as part of supervised work associated with the lectures of a basic didactic education unit of education sciences degree and master's courses at Aix-Marseille University in France in 2013 (and replicated annually since then). For the assessment of the course, students were required to present a didactic analysis of a didactic situation that students had to find themselves. The work required was divided into two steps. The students' first task was to carry out a search to find a (written or filmed) document presenting a didactic situation. The second task was a didactic analysis of the situation, encouraging the extension of a praxeological analysis of the didactic stake.

The first stage of the work quickly revealed that didactic situations were not easy to find. While at first glance the task seemed easy to the students (there would be an incalculable quantity of educational and "didactical" resources on the internet), the search quickly proved more demanding than expected. It was not, in fact, a question of finding documents that simply testified to the didactic intention of their authors, but of finding documents testifying to the presence of all the players in a *complete* didactic system. To guide them in the quest, a study of different types of situations was carried out during lectures and gave rise to the following specifications:

1. The first type of document is one that features a narration (by an author z) of a particular situation that has actually been observed—in which, for example, an instructor y attempts to have trainees X carry out a certain task t. Such a text could, therefore, be the narration (by z) of a training course carried out within a leisure-related or training institution. Note that in some cases we can have $y = x$: the document then describes person x attempting—using self-help—to carry out a certain task t.

2. The second type of document consists of accounts in which an author z, acting as a sort of instructor's instructor, describes to a hypothetical instructor y, a didactic situation involving hypothetical persons in training x, a situation which he/she would act to create (or which might quite simply happen). In this case, the user y referred to by the author of the text z is meant to extract from it a certain didactic structure to implement it in an institution with hypothetical persons x. Once more, we can have $y = x$ here.

3. The third type of document describes a certain praxeological structure presented as a possible didactic stake of a certain type of didactic or autodidactic system. In this case, the document does not describe a didactic situation either observed or to be created: it is a document related to a praxeological structure in which it acts as a possible study aid.

Regarding documents of the third type, we can give the example of a video in which a dentist explains "how to brush your teeth properly". In the video we can see only one person, y, talking to a supposed x, who is watching the video. This document typically presents an incomplete didactic situation. To move from a document of the third type to a document of the first type, it would be necessary for the dentist y in the video to explain to a young boy x, for example, how he should brush his teeth; and that we see on this occasion several interactions between y and x (for example, y commenting and correcting x's way of doing it, x questioning y, etc.). Failing that, we could only imagine what x might do: to achieve a didactic analysis of this type of document in a realistic way, you doubtless must make assumptions based on knowledge about the subject matter, which is not required at this level of didactic training.

With these instructions complicating the search for didactic situations, we will now see what our students performed through the praxeological analysis of their search accounts. The work to be carried out included a first section in which students had to explain how their search for a didactic situation panned out: (1) how the situation had been found (e.g. search on Google or a particular website; memory of a passage in a book or manual; a real situation filmed by themselves, etc.); (2) where the account of the situation had been found (e.g. a reference from a book or a video).

The analysis of the praxeologies students declared in this first section led to a series of observations and conjectures on two levels: first, the information search techniques used (the praxis level), and second, the discourse about these techniques (the level of the logos). The analysis also revealed the conditions and constraints affecting the performance of the type of tasks "Searching for a didactic situation of a certain type". Finally, the corpus revealed the full difficulty of accomplishing the type of tasks "Writing up the research process".

The main results of the study of the 89 student assignments we collected—65 projects by degree students and 24 projects by master's students—show that most students devoted little time to explaining their search process. This purely quantitative inadequacy is confirmed by the qualitative praxeological analysis that identifies the praxeologies listed by the students.

3.2.1 Poorly explained praxis

Types of tasks and techniques were not explained clearly in the students' assignments as we noted few specifications regarding the tools used to carry out the searches. Keyword searches, using the Google search engine or YouTube website, was the most widespread search technique: 39 students quoted it. The choice of keywords was specified in 30 cases, of which 22 named keywords specific to a particular field (dance, tennis, language, mathematics …) and eight used generic keywords. We can ask ourselves whether the account of the search was created a posteriori to meet the instruction to specify how the situation was found, the detailed search path, the keywords typed, and the links clicked on being certainly forgotten in many cases, whereas the instruction to list them was given at the beginning. We noted that some students took the trouble to write a series of keywords, while others opted for a vague expression, supposing that that would be sufficient to find the situation.

One thing appeared clearly when reading the accounts: they could only on rare occasions be used as a reproduction of the search carried out. They were certainly not drafted with the intention of acting as an operational model. The quasi-general lack of care given to their drafting can be seen as evidence either of a real difficulty in writing notes of this type or of a naturalisation of search techniques, the reporting of which would not require particular drafting. It was as if this type of tasks was a matter of course and that speaking about one's search was of no value in the work being carried out (some students even omitted writing this section). We found here the denial of problematicity studied in the ATD, which we had already observed in the study of internet information search practices, where there is a tendency to stick to tasks which do not a priori appear problematic while giving up as soon as the problematicity of the envisaged task becomes evident by subsequently "forgetting" such episodes (Ladage, 2008).

3.2.2 An incomplete logos

The students described a large number of difficulties in their statements, of which the following examples express a general perception:

> "After long hours of searching, and many websites visited, I found a didactic situation"; "The didactic situation presented here was painstakingly searched for"; "The search for a didactic situation was far from straightforward. One day, by dint of searching I ended up finding this video, but I no longer remember what I typed in to find it."

The difficulties are rarely explained and we find out little about the techniques used and even less about the justification for the techniques. The students' accounts do not allow us, in the great majority of cases, to understand precisely what was done, how it was done or why it was done in that way. The description of the searches is limited to the following simplified schema: I went onto the internet (or Google or YouTube) and I typed in some keywords. What happened then seems to have been a long and difficult exercise of which we know very little. The technical work that the students reported was only rarely accompanied by a technological explanation. Here we find what Chevallard (2012) calls a phenomenon of technological "muting": the technological discourse becomes inaudible; technology becomes silent. In other words: technological discourse is absent, be it voluntarily or habitual.

3.3 Performing didactic and praxeological analyses

The didactic analysis to be carried out had to include, as far as possible, an analysis based on the list of questions mentioned above. Students were also warned that, in responding to these questions, they might have to identify (and write down in the analysis to be submitted) the main "sets" of *conditions and constraints* which make possible, facilitate or on the contrary prevent (or, at least, hinder) the occurrence of such and such a *state* of the didactic systems examined. They were supposed to refer to the scale of levels of didactic codeterminacy studied on the course.

We cannot study in detail the type of analyses carried out by the students; we will, therefore, present four prominent traits, highlighting the conditions and constraints affecting the accomplishment of a didactic analysis within the context of the study of didactics at university.

3.3.1 A multiplicity of didactic stakes

A study of the didactic stakes identified by the students reveals a great disparity in subjects and levels of analytical depth. We provide a few examples below (the expressions are those chosen by the students):

> sign language / writing in French / expressing oneself correctly in French / expanding vocabulary in relation to kitchens and foodstuffs / the offset grip in basketball / developing phonological awareness / teaching spoken English to junior school pupils / performing the character of Phedra / learning the date in nursery school / Thales' Theorem / launching a mini-boomerang ...

Careful reading allows us to detect the difficulty some had in expressing the specific didactic stake of a given session to be measured. Thus, one often sees that the stake is defined at the level of the work as a whole, without the sections or subsections of the work having been identified.

3.3.2 Advanced analytical didactics

The list of didactic stakes presented above shows that, despite the apparent similitude of certain choices of situations related to the learning of the same work W, these rarely relate to the same section or subsection of the work. Confronting the choice of situation by each student on a shared online course forum highlighted a variety of partial didactic stakes, which could all contribute to the more general study of a work W.

3.3.3 A certain pervasiveness of the retrocognitive syndrome

Although students on the course were strongly advised not to look for a situation in a field familiar to them, some students admitted that they had been guided in their choice by their experience of or "passion" for the work studied. We had observed this behaviour among other groups of education science students during an investigation on sustainable development, where it could be observed that, "generally speaking", students only dealt with what they were supposed to know already: it is in this regard that we speak of a retrocognitive study mode (also referred to as retro-active study mode (Ladage & Chevallard, 2011, p. 335)).

3.3.4 An unrivalled depth of analysis

The group of master's students overall carried out more in-depth analyses than the degree students' group. It is important to specify that the educational conditions of the master's course greatly helped to engage the students in a more proactive study mode: their smaller number (24 students, as opposed to 65 in the degree group) promoted the use of supervisory didactic gestures during course time. These even led to collective searches on the internet, thanks to the fact that the students brought in their own computers.

3.4 Discussion

The results we obtained from this experiment revealed that embarking on the paradigm of "questioning the world" associated with a recognitive study mode would seem impossible at the present without obstacles or the construction of specific didactic conditions. Certain difficulties should be underlined. We should first note the pervasiveness of the paradigm of "visiting works" and that of a retrocognitive study mode, which means that students do not embark spontaneously on the search for a didactic situation when they are totally or almost totally ignorant of its subject. However, the pedagogy of the inquiry implemented in the context of an internet survey workshop with middle-school pupils (Ladage & Chevallard, 2011) testifies that it is possible to tackle apparently complex questions even with a young audience.

We should next note the use that the students declared they made of the internet, demonstrating their incapacity to effectively account for how they carried out their searches. This offers additional evidence of what we observed in 2011, namely a whole set of obstacles, specifically linked to entering the internet world in the dominant culture, which hinders the advent, dissemination and popularisation of a pedagogy of inquiry (Ladage & Chevallard, 2011).

A final difficulty that deserves to be highlighted concerns the time necessary for the teacher y to support students in carrying out their investigations and analyses. Supervisory didactic gestures carried out by y on the online course forum were certainly able to enrich the work of the students, but we are convinced that a more operational means has yet to be developed. It is a fact that teachers of education science students have the advantage of huge amounts of raw didactic material all around us, within society (and not only in "traditional" academic institutions). It is also true that the didactic tends to be hidden from the view of those uneducated in didactics and that it is therefore necessary to learn how to locate it, see it, identify it and, finally, fully analyse it in order to have any chance of improving its social efficacy.

4 General conclusion

What we have attempted with the description of the research carried out in the two domains studied in this chapter is to show the broad possibilities of the ATD framework as a body of knowledge useful for the research community outside mathematics, as well as for teachers and educators in general. Today there is plenty of room not only for empirical studies to confirm and further validate the theoretical value of ATD as a framework in general but also to experiment further with its research methodologies as well as its implementation in professional situations and training to develop and reinforce the dialectic between theory and practice.

References

Achiam, M., & Marandino, M. (2014). A framework for understanding the conditions of science representation and dissemination in museums. *Museum Management and Curatorship*, 29(1), 66–82.

Artigue, M., & Winsløw, C. (2010). International comparative studies on mathematics education: A viewpoint from the anthropological theory of didactics. *Recherches en Didactique des Mathématiques*, 30(1), 47–82.

Black, G. (2012). *Transforming museums in the twenty-first century*. London: Routledge.

Braun, V., & Clarke, V. (2006). Using thematic analysis in psychology. *Qualitative Research in Psychology*, 3(2), 77–101.

Chevallard, Y. (2011). Théorie anthropologique du didactique & ingénierie didactique du développement. *Journal du séminaire TAD/IDD*. Retrieved from http://yves.chevallard.free.fr/spip/spip/IMG/pdf/journal-tad-idd-20102011-7.pdf.

Chevallard, Y. (2012). Didactique fondamentale. Module 1: Leçons de didactique. Retrieved from http://yves.chevallard.free.fr/spip/spip/IMG/pdf/DFM_2011-2012_Module_1_LD_.pdf.

Hooper-Greenhill, E. (1999). *The educational role of the museum* (2nd ed.). London: Routledge.

ICOM. (2007). *ICOM Statutes*. Vienna: ICOM.

Kreps, C. (2003). Curatorship as social practice. *Curator: The Museum Journal*, 46(3), 311–323.

Kreps, C. (2006). Non-Western models of museums and curation in crosscultural perspective. In S. Macdonald (ed.), *A companion to museum studies* (pp. 457–472). Malden, MA: Blackwell Publishing.

Ladage, C. (2008). Étude sur l'écologie et l'économie des praxéologies de la recherche d'information sur internet : Une contribution à la didactique de l'enquête codisciplinaire (Doctoral dissertation). Université d'Aix-Marseille, France.

Ladage, C., & Chevallard, Y. (2011). Enquêter avec l'Internet: Études pour une didactique de l'enquête. *Éducation & Didactique*, 5(2), 85–116.

Marandino, M. (2016). The expositive discourse as pedagogical discourse: Studying recontextualization in the production of a science museum exhibition. *Cultural Studies of Science Education*, 11(2), 481–514.

Marandino, M., & Mortensen, M. (2011). Museographic transposition: Accomplishments and applications. In M. Bosch *et al.* (eds), *Un panorama de la TAD* (pp. 203–216). Barcelona, Spain: Centre de Recerca Matemàtica.

Marandino, M., Achiam, M., & Oliveira, A. D. (2015). The diorama as a means for biodiversity education. In S. D. Tunnicliffe & A. Scheersoi (eds), *Natural history dioramas: History, construction and educational role* (pp. 251–266). Dordrecht, The Netherlands: Springer.

McPherson, G. (2006). Public memories and private tastes: The shifting definitions of museums and their visitors in the UK. *Museum Management and Curatorship*, 21(1), 44–57.

Mortensen, M. F. (2010). Museographic transposition: The development of a museum exhibit on animal adaptations to darkness. *Éducation & Didactique*, 4(1), 119–137.

Mortensen, M. F. (2011). Analysis of the educational potential of a science museum learning environment: Visitors' experience with and understanding of an immersion exhibit. *International Journal of Science Education*, 33(4), 517–545.

Van Praët, M. (1989). Contradictions des musées d'histoire naturelle et évolution de leurs expositions. In B. Schiele (ed.), *Faire voir, faire savoir: La muséologie scientifique au présent* (pp. 25–34). Québec, Canada: Musée de la Civilisation.

PART 2

Praxeological and Didactic Analysis

3

PRAXEOLOGIES TO BE TAUGHT AND PRAXEOLOGIES FOR TEACHING

A delicate frontier

Michèle Artaud

1 From mathematical to didactic organisations and backwards

The theory of *praxeologies* (see Glossary), which includes praxeologies to be taught and praxeologies for teaching, was developed as an answer to the following question: where do personal and institutional *relations* to objects (see Glossary) come from? However, this theory is often used to analyse classroom activities or design didactic engineering proposals without considering their effects on personal or institutional relations. In this context, we run the risk of mixing both aspects because the frontier between praxeologies to be taught and praxeologies for teaching seems to be of importance to explain the formation of personal and institutional relations. This frontier should not be assumed as given: it depends on the didactic institution considered. We will see that it also depends on conditions related to the society and the civilisation level of the *scale of didactic codeterminacy* (see Glossary).

Let us consider first the effects of the institutional conditions and constraints. A number of praxeological analyses show that some praxeological organisations are *diffracted*: they involve a lot of types of tasks that are not connected to each other, and some of these types of tasks can be related to different techniques. This diffraction is an institutional construct that can be related to the *moment of the institutionalisation* (see Glossary), when decisions are made about what will be part of the knowledge organisation at stake and what will stay in the didactic organisation, remaining as a tool or canvas for the construction. In section 2 we will see an example of a class studying right-angled triangles and will highlight that the diffraction results from the integration of some ingredients of the didactic organisation into the mathematical organisation.

In the second example presented in section 3, we turn to examining how to integrate a didactic element, the *dialectic of media and milieus* (see Glossary), in a mathematical organisation. This dialectic is one of the nine dialectics identified by

Yves Chevallard to characterise knowing-through-inquiry didactic organisations (Kim, 2015). It starts from the fact that any assertion obtained from a media is a priori a conjecture and that its validity has to be tested by confronting it to some milieus. We will see two main ways of integrating this dialectic: by integrating new types of tasks for checking or controlling the mathematical work in the types of tasks of the mathematical organisation, or by including them as part of the initial techniques. In other words, one way consists of enlarging the type of tasks with new elements, the other in developing the initial techniques. We will see that the integration in the techniques leads to less diffracted praxeologies.

Of course, not all the didactic types of tasks and even less all the elements of didactic organisations can be integrated into the praxeologies to be taught. Distinguishing the two is methodologically essential to highlight didactic phenomena. Section 4 illustrates this aspect with a third example of a praxeological analysis of a mathematics session. The elements highlighted illustrate the risks researchers run while trying to identify the boundary between didactic organisations and praxeologies to be taught.

2 Institutionalisation and amalgamation

We are going to examine how a certain didactic praxeology related to the institutionalisation moment *diffracts* the mathematical organisation that is taught. For this purpose, let us consider the following episode observed in a class of 13–14-year-old students working on right-angled triangles and circumcircles. Different ingredients of the mathematical praxeology have emerged from previous activities and the teacher has proposed writing a summary as the work progresses. The summary, which describes some elements of the mathematical organisation, consists of four steps: "Showing that a point lies on a circle"; "Showing that a triangle is right-angled using the circumscribed circle"; "Calculating a length"; "Showing that a triangle is right-angled by looking at the median". Each step of this part of the institutionalisation moment follows the same pattern. First, the property that has emerged from the activity is written on the blackboard by the teacher and copied on the summary part of their notebook by the pupils. Then an example is treated by the pupils in collaboration with the teacher, which allows the institutionalisation of the practice regarded as the *didactic stake* (see Glossary).

In this summary, the mathematical organisation including the type of tasks T_{right} "Showing that a triangle is right-angled" appears to be made up of two subtypes of tasks that can be formulated as follows: "Showing that a triangle whose circumscribed circle is known is right-angled"; "Showing that a triangle in which the length of the median is known is right-angled". This type of synthesis leads to the integration of some *chronogenetic* elements related to the didactic organisation into the knowledge organisation. Indeed, each point-praxeology is institutionalised as far as it emerges and, therefore, we have the trace of the progress of the didactic time in the written traces of the synthesis.

The main type of tasks T_{right} that pupils have to solve consists of "showing that a triangle is right-angled" with whatever technique pupils wish to use, or can use, taking into account the information given in the problem. To be more functional and adapted, the mathematical organisation should then be *amalgamated* around the type of tasks "Determine *whether* a triangle is right-angled". The amalgamation of the mathematical organisation can be performed by creating a technique of the following type, which embraces the previous sub-types of tasks:

- If the length of each side of the triangle is known, compute the square of the longest side on the one hand and the sum of the squares of the two remaining sides on the other hand. If the two values are equal, then the triangle is right-angled; if they are unequal, the triangle is not right-angled.
- If the vertices of the triangle lie on the same circle, and if the diameter of the circle is the longest side of the triangle, then the triangle is right-angled; if not, the triangle is not right-angled.
- If [...]

In this case, the trace of the progress of the didactic time disappears from the mathematical organisation by the realisation of the amalgamation.

This type of amalgamation, performed at the level of the techniques, cannot be found in written syntheses today in France. The clinical observation of sessions shows that the oral institutionalisation in the classroom amalgamates types of tasks but leaves the techniques unrelated. Therefore, what emerges is an ecologically fragile local mathematical organisation instead of a fully functioning point mathematical organisation. Of course, teaching techniques to realise institutionalisation moments could be modified in order to improve this situation. This would require a modification of the praxeologies to be taught through a detailed analysis of what elements would be part of the techniques and what of the types of tasks—a detailed praxeological analysis that is not available to secondary school teachers today.

3 Integrating the dialectic of media and milieus into the knowledge to be taught

To continue the study of the integration of didactic elements into the praxeologies to be taught and its effects, we will consider here the case of the dialectic of media and milieus. As mentioned above, this dialectic considers that any assertion made by a media is a priori a conjecture: its validity has to be tested by confronting it with some milieus. The dialectic of media and milieus requires the performance of types of tasks such as T_{ch}, "Checking an assertion", or T_{co}, "Controlling a result". Even though these types of tasks are not specifically mathematical ones, we will focus on the mathematical context in what follows. The existence of tasks of type T_{ch} or T_{co} in didactic systems is difficult unless they are incorporated into the knowledge organisation.

Let us take the case of a local mathematical organisation about analytic geometry in the plane that could exist in France in grade 10 (15–16-year-old students). This local mathematical organisation contains two main types of tasks: T_{col}, "Determining whether or not three points are collinear" and T_{lin}, "Determining whether two lines are parallel or intersecting". T_{col} contains two subtypes of tasks: T_{eq},

"Determining the equation of a line [slope-intercept form]" and T_{pt}, "Determining whether a point lies on a line or not". Let us take T_{col} as the touchstone. Here is the institutional technique commonly used:

> τ_{col}: If necessary, choose a coordinate system and determine the coordinates of the three points, A, B, C. Determine the equation of the line (AB) and determine whether or not C lies on (AB).

To justify this technique, the technological-theoretical environment has to include at least the following results:

- Three points are collinear if they lie on the same line
- Two points define a line
- The equation of a line is $x = c$ or $y = ax + b$
- A point M $(x_M\ ;\ y_M)$ lies on the line of equation $y = ax + b$ (resp. $x = c$) if $y_M = ax_M + b$ (resp. $x_M = c$).

There are two ways of integrating the dialectic of media and milieus in a knowledge organisation: at the level of the types of tasks (the T-way) and at the level of the techniques (the τ-way). Let us consider the T-way first. It consists of the addition of four types of tasks specifying T_{ch} to our local mathematical organisation: T_{ch_col}, "Checking whether three points are collinear or not"; T_{ch_lin}, "Checking whether two lines are parallel or intersecting"; T_{ch_eq}, "Checking the equation of a line [slope-intercept form]"; T_{ch_pt}, "Checking whether a point lies on a line or not". We then have to add four techniques, one for each type of tasks. Here is, for example, a technique for T_{ch_col}:

> τ_{ch_col}: Plot the three points on a graph; draw the line passing through two points and examine whether the third one lies on this line.

Let us now consider the integration of the dialectic of media and milieus in the τ-way. We keep the four types of tasks and modify the technique of each one by including some subtypes of tasks of T_{ch} or T_{co} *within* the technique. It is therefore a different strategy from the previous one where we work on the techniques while keeping the types of tasks. Let us take T_{col} as the touchstone another time. The technique τ_{col} can be modified as follows:

> τ_{col_m}: If necessary, choose a coordinate system and determine the coordinates of the three points, A, B, C. Determine the equation of the line (AB) and determine whether C lies on (AB). Check on a figure, drawing line (AB) and examining whether C lies on (AB). Check coordinates of the points and check the equation of the line.

The technological-theoretical environment must be completed with some additions, such as justifying that results have to be verified.

If we compare the two ways of integrating the dialectic of media and milieus in the praxeology to be taught, we can stress that, on the one hand, with the *T*-way, the mathematical organisation is not much integrated: it now includes some "checking tasks" that are difficult to make exist and unlikely to be identified by the students. On the other hand, with the τ-way, the mathematical organisation makes the functions of the dialectic of media and milieus more evident, and then the dialectic of media and milieus increases the reliability of the techniques. Moreover, in the τ-way, the dialectic of media and milieus is included in the students' *topos* (see Glossary): they are responsible for carrying it out. This can favour the building up of the mathematical organisation. However, the fact that checking the validity of the tasks becomes part of the students' responsibility goes beyond a simple modification in the conditions and constraints relating to the sector or even the domain level in the scale of codeterminacy. Installing the dialectic of media and milieus in didactic systems requires changes at the higher levels of the scale—pedagogy, school and, first of all, society. This point will not be developed here.

We have seen that some pieces of the didactic organisation can be included in the knowledge organisation (as in the above example, through the conditional tests "if...") and that this seems better achieved through the τ-way. It yields a fully functional knowledge organisation and extends the students' *topos*. However, not everything can be included in the praxeology to be taught: some didactic entities, involved in the realisation of didactic moments, have to remain as parts of the didactic organisation related to the didactic stake. We are now giving an example of this assertion.

4 Mathematics included in the didactic organisation

For this example, we will analyse a 55-minute session observed by a teacher educator in a class of 13–14-year-old students. The session was the second in the teaching sequence, and we have an observation report made by the teacher educator. In the first session, the class had studied the type of tasks "determine the length of one side of a right-angled triangle, adjacent to an angle of 60°". The students found that the ratio of the adjacent side to the hypotenuse was constant and was 0.5 in this case. This allowed them to formulate a conjecture: there is a relationship between the angle and the ratio between the adjacent side and the hypotenuse. Then they proceeded to test this conjecture in the case of a 70° angle. The main purpose of the second session is to test this conjecture with the dynamic geometry software Geogebra.

Students construct a right-angled triangle, measure one of the acute angles, calculate the ratio of the side adjacent to this angle and the hypotenuse, and check that it is constant by varying the measures of the sides of the triangle (they must obtain at least five values of the sides). The work carried out leads to the following statement written on the blackboard and in notebooks: "In a right-angled triangle, for a given acute angle, there is a link between the measure of the angle and the quotient of the measure of the adjacent side to the measure of the hypotenuse." The quotients found experimentally are recalled, the teacher states that the common value of the ratio is called cosine of an angle and distributes a cosine table. Students check that the values they have found are consistent with those of the table. A sheet showing three "applications" is finally distributed; here are the statements:

> Application 1: ABC is a right-angled triangle in B. We know that $\angle BAC = 60°$ and $AB = 12$ cm. What is the measure of the length of segment $[AC]$?
>
> Application 2: EFG is a right-angled triangle in G. It is known that $\angle EFG = 45°$ and $EF = 10$ cm. What is the measure of the length of segment $[FG]$? (Round to nearest 0.1.)
>
> Application 3: IJK is a right-angled triangle in I. It is known that $JK = 5$ cm and $IJ = 4$ cm. What is the measure, rounded at the unit, of angle $\angle IJK$?

The two first tasks are to be carried out for the following session, and the few remaining minutes of the observed session are devoted to a collective explanation of the work to be done for the resolution of application 1.

A first analysis could reveal T_{rat}, "Determining in a given right-angled triangle the ratio between the adjacent side of an acute angle and the hypotenuse", as a type of tasks forming part of the mathematical organisation that is the didactic stake. It is clear however that the purpose of the study of the session is to bring out the notion of cosine of an angle. This is the main element of the local mathematical organisation at stake, with the two types of tasks that appear in the exercises given at the end of the session: "Determining the measure of the length of a side of a right-angled triangle" and "Determining the measure of an angle of a right-angled triangle". The T_{rat} type of tasks is, therefore, an ingredient of the *didactic* organisation: it participates in the rea-lisation of the technological-theoretical moment, justifying the existence of a constant ratio that depends only on the angle between the adjacent side of one of the acute angles of a right-angled triangle and its hypotenuse, and producing the values of this ratio for some given measurements of this angle.

Other elements corroborate the above analysis, including the fact that, as the report of the session clearly shows, the T_{rat} type of tasks is not a problematic type of tasks at the school level under consideration and is therefore likely to be part of the milieu of the study in progress.

This type of misunderstanding—including a didactic type of tasks in the mathe-matical organisation—masks the implementation of the technological-theoretical moment. Indeed, when producing proofs or elaborating definitions is not seen as a part of the realisation of this moment, the fact of justifying, producing or making intelligible the techniques appears as an undeveloped, even prevented action.

In particular, in the observed session, it is not clear whether the calculations of quotients of adjacent sides to hypotenuses are meant to "calculate cosines" or to establish the existence of a numerical invariant attached to an angle and then that the cosine is well-defined. This can lead to forgetting the production work of the proofs or definitions in favour of the exhibition of a beforehand shaped demon-stration or definition, which is part of a type of realisation of the institutionalisation moment. Here, the teacher's institutional relation to mathematics could be poorly adapted. For the researcher, this leads at the least to not being able to see many phenomena related to the realisation of these two didactic moments.

The situation is a little different for the type of tasks "Determine the cosine value of an acute angle in a right-angled triangle" that one might want to integrate into the mathematical organisation. The observed session highlights that this type of tasks will not be isolated but integrated into the technique of length calculation or angle determination. The questions asked through the exercises at the end of the session attest to this, and this type of tasks only existed in isolation for a relatively short period of time before being amalgamated with and integrated into the techniques.

The second misunderstanding refers to the diffraction of mathematical organisations in praxeological analysis. On the one hand, if the praxeological analysis is used to build a mathematical organisation to be taught, a maladjusted institutional relation in the student's position is created because it consists of numerous juxtaposed point-praxeologies, such as "determining the cosine value of an acute angle in a right-angled triangle", which could be enhanced by amalgamation at the τ-level. On the other hand, in the conduct of the study of this mathematical organisation, some *topos* is removed from the students since, for instance, they did not have to identify by themselves the subtype of tasks about length calculation they have to accomplish.

5 Fighting the denial of the didactic

We can see that integrating some elements of the didactic organisation in praxeologies to be taught can diffract these praxeologies, and this diffraction can lead to difficulties, especially in the construction or the analysis of the students' relation to the praxeology to be taught. Yet, one can wish to integrate some elements of the didactic organisation into the praxeologies to be taught because it is difficult for specific didactic types of tasks to exist, even in didactic institutions. The reasons are to be found at higher levels of the scale of didactic codeterminacy (Figure 3.1; Chevallard 2013). In particular, the constraint imposed by the *denial of the didactic*, which we put at the civilization level, is important. This constraint is illustrated by the fact that cultural and social discourses remain almost silent about what has to be learnt and what one has to do to learn it.

The previous analysis of the session on the cosine gives an example of one of the manifestations of the *denial of the didactic*, particularly at school and discipline levels. It consists of the attribution to knowledge of elements that are properly linked to the study process and that the praxeological analysis can be brought to light (Artaud, 2011). This problem of highlighting the frontier between what is at stake

---

Humankind ⇆ Civilisation ⇆ Society ⇆ School ⇆ Pedagogy ⇆ Didactic System ⇆ Discipline(s) ⇆ Domain ⇆ Sector ⇆ Theme ⇆ Subject

---

FIGURE 3.1 The scale of levels of didactic codeterminacy

in the study process—praxeologies to be taught—and what constitutes a means for the study—praxeologies for teaching—also appears in a more underground way in research work of phenomenotechnical scope. For instance, when developing a didactic engineering proposal, it is not always easy to distinguish between what will constitute the student's expected answer and what participates in its elaboration. It seems important to develop research praxeologies so that the distinction between praxeologies to be taught and praxeologies for teaching is taken into consideration: it is the price to pay for being able to identify the effects of the denial of the didactic and analyse how this constraint works.

References

Artaud, M. (2011). Les moments de l'étude: Un point d'arrêt de la diffusion ? In M. Bosch *et al.* (eds), *Un panorama de la TAD* (pp. 141–162). Barcelona, Spain: Centre de Recerca Matemàtica.

Chevallard, Y. (2013). Fondements et méthodes de la didactique des mathématiques. Enseignement pour le parcours "Didactique" du master "Mathématiques et applications" de l'université d'Aix-Marseille. http://yves.chevallard.free.fr/spip/spip/article.php3?id_article=219.

Kim, S. (2015). Les besoins mathématiques des non-mathématiciens : quel destin institutionnel et social ? Études d'écologie et d'économie didactiques des connaissances mathématiques (Doctoral dissertation). Aix-Marseille Université, France.

4

DEVELOPMENTS AND FUNCTIONALITIES IN THE PRAXEOLOGICAL MODEL

Hamid Chaachoua, Annie Bessot,
Avenilde Romo and Corine Castela

This chapter presents two main developments of the praxeological model. In relation to the theory of didactical situations, the first sections introduce the two notions of variable and generator of a type of tasks. These notions make it possible to structure a set of types of tasks to build a reference praxeological model on elementary algebra. The second development focuses on the institutional determination of praxeologies, with specific attention to the interactions between technologies and institutional technical standards as well as to transposition phenomena.

1 Introduction

The analysis of knowledge in terms of praxeologies aims at better capturing the specificities of the knowledge as it is proposed to be taught as well as the knowledge that is actually taught or learnt. The notion of *praxeology* (see Glossary) is essential for researchers to detach from the school conception of knowledge and elaborate their own reconstructions of what appears to be a central element of teaching and learning processes. It also contributes to overcoming the division between knowledge products and the process of elaborating, adapting, using or simply studying these products. The first example of such an analysis (sections 1–4) proposes to include the notions of *variable* and of *generator type of tasks* in the praxis block of praxeologies. This extension is required when using the praxeological analysis in technology-enhanced learning (TEL), especially to elaborate learning trajectories and provide automatic feedback to both teachers and students. The second example (sections 5–7) focuses on the logos block of praxeologies and on the transformations operated in it during transposition processes, when a praxeology is moved from the institution where it is produced to the one where it is used, and from both of them to the institution where it is taught for future users.

2 The notion of variable in the praxeological model

One of the grounded ideas of the theory of didactical situations (TDS, Brousseau, 2002) is modelling knowledge through situations. The fundamental methodological principle proposed by Guy Brousseau is to define any piece of knowledge by a "situation", i.e. by a game against a *milieu* that models the problems this piece of knowledge can optimally solve. Situations are minimum models for studying the role of knowledge in the particular relations between a person and a milieu. The notion of variable appears intimately intertwined with the notion of situation:

> [Situations] must have variants and variables. It is then possible to investigate by mathematical and experimental means, which values of these variables can determine the optimal conditions for the diffusion of a determined piece of knowledge, or explain those that appear as the (theoretically) optimal response to the conditions proposed to the student. For given values of these variables, there is at least one optimal strategy (in terms of implementation cost, reliability, learning cost, etc.) and one or more corresponding pieces of knowledge. We call cognitive a variable of the situation such that by choosing different values we can bring about changes in the optimal piece of knowledge. The didactic variables will be among the cognitive variables those that can be set by the teacher.
>
> *(Brousseau, 1994, pp. 3–4, our translation)*

The introduction of variables aims to structure a set of specific situations of a piece of knowledge and to make this modelling computable. From this point of view, a set of specific situations of a given piece of knowledge will be characterised by a restricted set of pertinent variables. The notion of variable initially appears as a methodological instrument within a modelling process, associated with an a priori analysis of a particular or fundamental situation.

This notion allows consider a computable universe of variations of a situation linked to a piece of knowledge with a multiple purpose:

- Generating a field of mathematical problems and practices considered as specific of the piece of knowledge
- Distinguishing different meanings of the piece of knowledge through situations that are different from a didactic point of view but equivalent according to the knowledge involved
- Producing experimental geneses of a piece of knowledge (didactic engineering)
- Studying the existent conditions of a piece of knowledge within a certain educational reality and the reasons for the observed difficulties.

The fundamental methodological principle mentioned before can be completed as follows:

It is then necessary to determine the set of situations that are likely to make a notion work, conferring on it the different meanings that determine the corresponding concept. Only the differences in situations that affect the concept are in the field of didactics; they are the result of variables to be determined in each case.

(Brousseau, 1981, p. 109, our translation)

The introduction of the notion of variable in the ATD is linked to research studies developed within the MeTAH team (Models and Technologies for Human Learning) on the design, development and use of technology-enhanced learning (TEL). The MeTAH team began to work on the computational modelling of the objects of knowledge that are to be taught, of the subject's knowledge, of the propositions for equipping the computational environments with calculation means for producing a diagnosis, of the feedbacks for the student or for the teacher, etc. This necessity led us to develop the reference framework T4TEL (T4 refers to the praxeological quadruplet type of tasks/technique/technology/theory and TEL for technology-enhanced learning).

T4TEL is in line with the praxeological approach (Bosch & Chevallard, 1999). It represents a formalisation and an extension of the praxeological model to answer a twofold need: from the one side, to have a model that is computable from a computational point of view and, from the other side, to produce different TEL applications. This introduction is also interesting from a didactic perspective; besides the integration of TEL, we find some common points with the notion of variable within the TDS.

3 The praxeological model T4TEL

3.1 The definition of type of tasks and of subtypes of tasks

Given a type of tasks T and a technique τ to perform them, we can consider a mathematical point-praxeology that brings together the tasks that can be accomplished by τ, justified by a technology θ that is legitimated by a certain theory $\ominus$ Nevertheless, we can have several techniques for the same type of tasks. Teaching institutions organise the study of these techniques but, as Yves Chevallard (1999) highlights, we generally limit the number of techniques for a certain educational level:

In a given institution I, for a given type of tasks T, there generally only exists one technique, or at least *a small number* of techniques *institutionally recognised*, excluding other possible alternative techniques—which can effectively exist, but in *other* institutions.

(Chevallard, 1999, p. 225, our translation)

Which are the conditions and constraints reducing the range of the techniques within a certain institution?

In the T4TEL model, a type of tasks is defined by an action verb and some complements. Within a teaching institution, we can affirm that the types of tasks included in the knowledge to be taught always contain at least one technique to carry them out. T4TEL introduces a more complex definition of the notion of type of tasks, based on the notion of technique's scope, noted $P(\tau)$. $P(\tau)$ is the set of tasks the technique τ allows to achieve under certain institutional conditions that are assumed to be stable over a certain period. To be considered as a type of tasks, a set of tasks T must contain at least a task t that can be achieved by a technique τ such as either all tasks achieved by τ belong to T or all tasks belonging to T can be achieved by τ. This can be stated as: A set of tasks T is a type of tasks if the following condition holds:

$\exists\ t \in T$, $\exists\ \tau$ achieving t, such as $P(\tau) \subseteq T$ or $T \subseteq P(\tau)$.

This definition says that a set of tasks may be defined as a type of tasks, even though no one technique "solves" all the tasks.

The notion of *subtype of tasks* features the case of two types of tasks, T_a and T_b with T_a included in T_b. Let us draw attention to the fact that a subset of tasks may not necessarily be considered as a subtype of tasks. The conditions of the definition of type of tasks apply.

For example, let us consider the type of tasks *Tra_eq2*: "Solving a quadratic equation" in secondary education. There exists a dominant technique that accomplishes some tasks of *Tra_eq2*: that of the discriminate. Other techniques also exist, such as the one based on the use of the square root allowing the accomplishment of the tasks "solving equations of the type $P_1(x)^2 = k\ (k > 0)$" (we indicate with $P_i(x)$ a polynomial the canonical form of which is of degree i. For example, $P_1(x) = (x + 1)^2 - x$ because its canonical form, $P_1(x) = 2x + 1$, is of degree 1). Hence, we can consider the set of these tasks as a subtype of tasks of *Tra_eq2*.

3.2 The description of the techniques

The problem of the description of the techniques was already mentioned by Marianna Bosch and Yves Chevallard (1999): what are techniques made of? What "ingredients" compose them? And also: what does the implementation of a technique consist of? Even if this problem is not explicitly considered, some ATD research studies describe techniques in the form of more or less structured actions, while others decompose them into subtasks. For example, Gisèle Cirade and Yves Matheron (1998) describe the technique used for the type of tasks "solving a first-degree equation" through the subtasks: developing an algebraic expression, calculating the products, transposing the terms, simplifying each member, solving an equation of the type $ax = b$. These authors add that this listing is related to a given educational institution and should be considered as a model to highlight the mathematical organisation that will be assessed later on. This listing is interesting because it refers to the types of tasks institutionally identified for which a mathematical praxeology is put into play. Consequently, this listing allows didacticians to consider students' difficulties in the implementation of a technique at the level of the "subtasks" that compose it. However, by doing this, we are not including the parts of the techniques that are not explicit and therefore cannot be mentioned in the considered institution.

Within T4TEL, a technique τ is described by a set of types of tasks {Ti} that can be of two types: (1) types of tasks only existing through the implementation of the techniques of certain other types of tasks, called *types of intrinsic tasks*; (2) types of tasks that can be institutionally prescribed to students, named *types of extrinsic tasks*.

For example, in France, at the end of primary school, within the technique of adding two whole numbers, there exists at least two types of tasks: "Putting the two numbers in a column" and "Adding two whole numbers between 0 and 9". The first type of tasks, "Putting two numbers in column", is intrinsic: it is not prescribed by the institution even if it has a praxeology and, in particular, a technique and technology. The second type of tasks, "Adding two whole numbers between 0 and 9", is extrinsic because it is prescribed at the beginning of primary school.

Finally, it should be noted that the ingredients of a technique are conditions of its existence. We will develop later this aspect in section 5.

3.3 The notions of variable and generator within T4TEL

We have already seen that within T4TEL a type of tasks is defined by an action verb and some complements. The action verb characterises the kind of task, such as "Calculating" or "Counting". The complements can be defined according to different levels of granularity, from the most specific to the most generic. For example, "Calculating the sum of two integers" is more general than "Calculating the sum of two one-digit integers".

How can we grasp the relations between the specific and the generic within the organisation of the types of tasks? More generally, and considering the notion of variable in TDS, how can we structure a set of types of task related to an object of knowledge that has to be taught and make this structuration "calculable"? To organise and structure the praxeology of a discipline or a discipline domain, we propose to rely on the techniques' scope.

Let T be a type of tasks and τ a technique of T with scope P(τ). From the analysis of school knowledge, we can advance the following remarks about P(τ) concerning T:

1. If T is included in P(τ), then it can exist a type of tasks T' more generic than T, to which τ is also a technique.
2. If P(τ) is included in T, then it can exist a more specific type of tasks T'' where τ is a technique.

We will now introduce the notions of *variable* and *generator* of a type of tasks to consider these relations and make them calculable. A generator of a type of tasks is defined by a type of tasks and a *system of variables*. We speak of a system of variables to refer to a list of variables together with certain values they can take. For example, the type of tasks "Calculating the sum of two numbers" defines a generator of the type of tasks GT: [Calculating the total of two numbers; V1, V2, V3], where V1 is the nature of the numbers (whole numbers, integers, decimals, etc.), V2 is the size (number of digits) of the first number, and V3 is the size of the second number. Other variables can, of course, be considered, as the ways of presenting the numbers, their representation, etc. The variables of a generator take values within a discipline domain. These values generate types of tasks more specific than the type of tasks T.

The first function of a variable is to generate subtypes of tasks playing on the values of this variable. For example, T_1 "calculating the sum of two whole numbers" or T_2 "calculating the sum of a whole number whose size is 1 and another one whose size is 3". T_1 and T_2 are subtypes of T, where T_2 is a subtype of T_1. These task subtypes can have specific techniques: for example, an economical technique of T_2 starting with the biggest number and "overcounting" the second number of units. This technique is less pertinent for two numbers of size 2 and 3. The second function of a variable is to allow the characterising of the techniques' scopes.

These two functions are essential in the a priori analyses—from an epistemological and didactic point of view—and for calculating learning trajectories starting from a change on the variables and on their values. In particular, the construction of a *reference praxeological model* (RPM) for a given mathematical domain involves a specific clarification of the variables and of their possible values. When using the values of variables of a generator of a type of tasks, we can distinguish a triple point of view: the epistemological, the institutional and the didactic.

We talk about *epistemological* variables when variables of a type of tasks are considered within a given RPM. In this case, the listing of the values of a variable is such that the change of a value modifies the range of the possible techniques of a type of tasks and, consequently, the relation to an object of knowledge.

Within an institution, there always exist conditions and constraints that reduce not only the type of tasks but also the possible values of the corresponding variables. For example, in the first level of French primary school, for the type of tasks T "Calculating the sum of two numbers", the numbers are always whole numbers (V1) and both numbers are less than 30 (V2 and V3). A variable and its institutional values model explicit or implicit conditions and constraints under which a praxeology exists or can exist in a given institution. The institutional listing of the values of a variable does not always have an epistemological pertinence.

A *didactic* variable is a variable within an institution that is potentially at the disposition of the teacher. A didactic variable within an institution and at a given moment does not need to be a didactic variable at a different moment or within another institution.

We can enrich the a posteriori values of didactic variables with additional values after considering the students' personal praxeologies and their validity (Chaachoua, 2010; Croset & Chaachoua, 2016). This is a third function of the notion of variable that is particularly important for the diagnosis of students learning trajectories within a given institution.

4 An example of generator of type of tasks

In this example we present a way to structure point-praxeologies attached to the generator of a type of tasks T "Solving an equation" integrated into the same *local* mathematical organisation. The diversity of techniques and the forms of the equations leads to associate a set of variables to the complement "equation". This work needs the construction of an RPM about this type of tasks. The construction of this reference model relies on the research studies of Croset (2009), Chaachoua (2010) and Pilet (2012) about the teaching of algebra in compulsory secondary education in France (11–16 years).

Within this RPM we distinguish a global mathematical organisation (GMO) of the algebraic domain starting from three regional mathematical organisations (RMO): RMO1 "Algebraic expressions", RMO2 "Formulas" and RMO3 "Equations". RMO3 is structured in different local mathematical organisations (LMO), one of which is LMO: "Solving equations". The type of tasks T "Solving an equation" is part of this local mathematical organisation that groups many point-praxeologies. We can define a generator of T by considering a system of variables {V1, V2, V3, V4}:

GT = [Solving an equation; V1, V2, V3, V4]

where V1, V2, V3, V4 are:

V1: Equation degree

In the case under study (compulsory secondary education in France), we consider three pertinent values (0, 1 and 2) that are connected to a first institutional constraint.

V2: Number of solutions of the equation

It can take four values: 0, 1, 2, infinity.

V3: Algebraic form of the left member

It can take different values: 0, positive constant not equal to 0, negative constant not equal to 0, first-degree polynomial, square of a first-degree polynomial, product of two first-degree polynomials, second-degree polynomial (and the possible cases of the polynomials according to the values of their coefficients).

V4: Algebraic form of the right member

It takes the same values as V3.

The instantiation of values of one or several variables of GT generates a structured set of subtypes of tasks. For example, the subtype of tasks "solving a zero-product equation of the type $P_1(x)Q_1(x) = 0$" is formally represented by $T_2 =$ (Solving an equation; V1 = 2, V3 = product of two first-degree polynomials, V4 = 0). This formalisation indicates that the type of tasks T_2 is generated by the generator of type of tasks GT. The absence of V2 indicates that all the values are possible. We highlight that variables are not independent and some combinations of values are not possible such as (Solving an equation; V1 = 1, V2 = 2, V3 = product of two first-degree polynomials, V4 = 0). Then, the values of some variables can depend on the values of other variables.

The variables values can be ranked. For example, for V3 they can be structured as shown in Figure 4.1.

If we fix the values of the other variables, the above hierarchy implicates a hierarchy of the associated types of tasks. For example, the type of tasks $T_1 =$ (Solving an equation; V1 = 2, V3 = second degree polynomial, V4 = constant) is more general than the type of tasks $T_3 =$ (Solving an equation; V1 = 2, V3 = second degree polynomial, V4 = positive constant not equal to 0). Then, T_3 is a subtype of tasks of T_1, and if a technique accomplishes the tasks of T_1 then it also accomplishes the tasks of T_3.

In the same way, if we consider the constant as a polynomial whose degree is 0, we obtain the structuring shown in Figure 4.2.

Some of the generated subtypes of tasks cannot exist within a given institution because of the institutional constraints of the variable values. For example, the sub-type of tasks $P_1(x)Q_1(x) = k$, $k \neq 0$ does not appear at the level of secondary school teaching in France. This could have consequences for the students' praxeologies, as shown below.

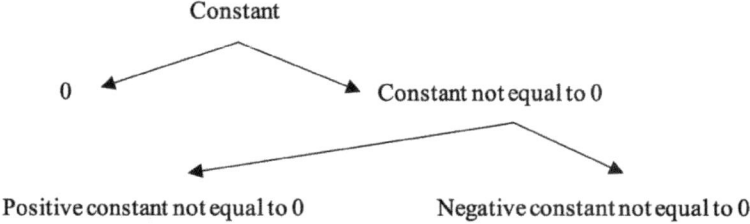

FIGURE 4.1 Example of a hierarchic tree for a subset of the values of a variable

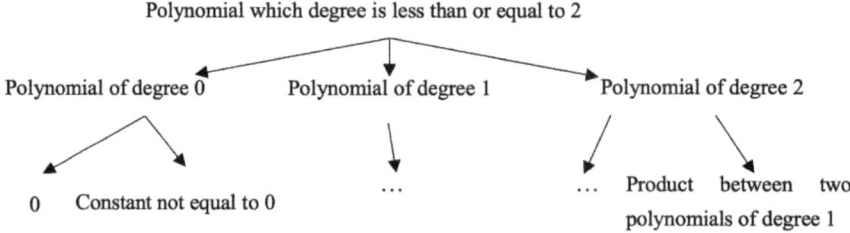

FIGURE 4.2 Example of a hierarchic tree for the set of values of a variable

5 Analysing students' personal praxeologies

Until now we have only considered institutional praxeological organisations. However, learning processes usually result in the students' production of praxeologies that differ from the institutional ones and can be more or less stable during a given period. Some of these praxeologies are based on the implementation of techniques that could be scientifically valid but institutionally inadequate, or just not scientifically valid.

Let us consider an example in relation to the generator of the type of tasks GT = [Solving an equation; V1, V2, V3, V4]. Let us consider the task "solving the equation $x^2 + 4x = -4$". The teacher waited for the following solution:

$$x + 4x = -4$$

$$x^2 + 4x + 4 = 0$$

$$(x + 2)^2 = x$$

$$x + 2 = 0$$

$$x = -2$$

This answer is the implementation of the following expected technique constituted by three types of tasks:

Grouping all of the terms in the left member
Factorising $P_2(x)$ through the technique of "the notable products"
Solving a first-degree equation of the type $Q_1(x) = 0$

Here below is the resolution of a grade 10 student:

$x^2 + 4x = -4$
$x(x + 4) = -4$
$x = -4$ or $x + 4 = -4$
$x = -4$ or $x = -8$

This student's resolution can be interpreted as the implementation of the following technique:

Factorising the left member
Solving an equation of the type $P_1(x)Q_1(x) = k$, $k \neq 0$
Solving two first-degree equations $P_1(x) = k$ and $Q_1(x) = k$

At the second resolution step, the student is faced with the non-institutional task "Solving the equation $x(x + 4) = -4$". She then adapts the technique she has learnt for the institutional type of tasks "Solving a first grade equation of the type $Q_1(x) = 0$" for this non-institutional type of tasks "Solving an equation of the type $P_1(x)Q_1(x) = k$, $k \neq 0$", taking *personal* responsibility. The combination of the values of V3 "product between two first-degree polynomials" and those of V4 "constant not equal to 0" allows grasping the personal praxeology of the type of tasks (solving the equation $P_1(x) Q_1(x) = k$, $k \neq 0$). This type of tasks can be considered as generated by GT and can be modelled by: $T_2' =$ (Solving an equation; V1 = 2, V2 = 2, V3 = product between two first-degree polynomials, V4 = constant not equal to 0).

The quadruplet of the personal praxeology of this student can be described according to the T4TEL model as follows:

> $T =$ (Solving an equation and V1 = 2, V2 = 2, V3 = product between two first-degree polynomials, V4 = constant not equal to 0)
> $\tau = \{$(Writing $P_1(x) = k$ or $Q_1(x) = k$); (Solving a first-degree equation)$\}$
> $\theta = \{$"If $P_1(x)Q_1(x) = k$, then $P_1(x) = k$ or $Q_1(x) = k$"; technological elements of the technique of the resolution of the first-degree equations$\}$.

We classically interpret the student's action of writing "$P_1(x) = k$ or $Q_1(x) = k$" as the application of a rule R outside its validity domain that corresponds to the case $k = 0$. This is the T4TEL model integrates this interpretation through the formulation "Applying the rule R", giving the following description of the personal praxeology within the T4TEL model:

> $T =$ (Solving an equation and V1 = 2, V2 = 2, V3 = product between two first-degree polynomials, V4 = constant not equal to 0)
> $\tau = \{$Applying the R rule; Solving an equation and V1 = 1$\}$
> $\theta =$ (R: "If $P_1(x)Q_1(x) = k$, then $P_1(x) = k$ or $Q_1(x) = k$"; technological elements of the technique of the resolution of the first grade equations).

To sum up, we can say that in the T4TEL model, the inclusion of variables in the praxeological analysis contributes to a structuration of praxeologies based on the structuration of variable values (model calculability). It also provides elements to characterise the institutional conditions and constraints for a given collection of praxeologies, together with the possible choices for the guidance of the study (didactic variables). Finally, by considering students' personal praxeologies, it can provide a detailed description of the learning trajectories, including types of tasks and techniques that are not always visible to the institution.

This formalisation has been applied to analyse students' personal praxeologies in algebra (Croset, 2009) and the school praxeologies of the perspective representation in the geometry in the space (Tang, 2014). The T4TEL model is currently being used to produce different EIAH services (diagnosis of knowledge, feedback for a subject within an institution) that are not exclusively based on the learner but take into account the idiosyncrasy of the considered institution. In particular, it has been used to model the domain of "number systems" at the end of primary school in France (10–12 years old).

It is the basis of some instruments for primary school teachers, like the characterisation of students' profiles or the production of tests of diagnosis focusing on the variables.

6 Praxeologies as institutional idiosyncrasies

As we have seen with the application of the T4TEL model, the ATD considers that human activities are institutionally situated. An institution I is a stable social organisation that offers a framework in which some different groups of people carry out different groups of activities. These activities are subjected to a set of conditions and constraints. Some of them are universal, others are institutionally specific, derived from the expectations of the institution I, themselves partly determined by a network of institutions influencing I. Hence, the ATD assumes that activities are institutionally contrived and, consequently, that the praxeologies related with these activities within I are impacted by the institutional influence. Chevallard (2006, p. 23) considers praxeologies as social idiosyncrasies.

To exemplify the fact that the institutional context regulates the praxeologies, let us consider briefly the case of nurses' medication dose calculations in a hospital setting (Roditi, 2014). A nurse has to administrate 540 mg of a prescribed drug to a patient. The drug is supplied in ampoules of 20 ml/250 mg. Two ampoules give 500 mg, so we need 40 mg more. To calculate the volume to remove from the third ampoule, which is a proportionality task, a usual technique is the cross product (here, $\frac{40 \times 20}{250}$). Yet Roditi's investigation has shown that in some hospital services, nurses do not use it in such a situation. Instead, they will first dilute one ampoule content with 5 ml of a neutral serum to obtain a solution of 25 ml/250 mg, so that they get a drug concentration of 10 mg/ml. Then the volume calculation will be very easy. Why is this technique promoted? The dose must be mentally computed for reasons of rapidity; moreover, mistakes must be avoided because they may be detrimental to the quality of the treatment, sometimes even fatal. Consequently, in such institutions, the dose calculation technique is expected to not only provide the correct result when properly implemented but also be reliable, in the sense that individual mistakes should be highly unlikely. We note that this technique is made possible by the availability in the institution of neutral physiological serum.

In general, constraints of different natures govern the praxis, depending on the institution's resources, needs and values. The technique will be expected to comply with specific standards which may relate to:

- its validity, with criterions regarding, not only its effectiveness (completing the task), but also its efficiency (reliability, generality, etc.);
- its usability and safety;
- its rational intelligibility;
- its adequacy to dissemination and learning.

This list of technical standards is not exhaustive. We will assume that, to be admitted in a given institution, a technique must be submitted to a process in which its adequacy in terms of the institutional standards is evaluated. This is the meaning we give to the idea of legitimation, referring to the Latin etymology (*lex*: law; *legitimus*: in conformity with the law). The modes of legitimation depend on the institution, on the technique and the controlled standard. For instance, to validate a mathematical technique τ, a mathematics research institution produces a proof based on some theorems. This proof belongs to the technology of τ and is in turn supported by a mathematical theory. We know that this legitimating mode is a mathematical idiosyncrasy; generally, the effectiveness of a technique is controlled through devices and procedures that are not purely discursive. The technology of the technique related to the validity may include a description of these devices and procedures (e.g. presentation of the laboratory protocol in experimental sciences). However, in some cases, especially when the control is empirical, more or less informal, based on repeated use by numerous users, the technology may merely formulate the guarantee that the technique is considered as effective according to institutional criteria. Regarding mathematics, this may be the case for technological elements related to the technique's efficiency and usability (what is the technique's scope? Which errors are more frequent? How to avoid them?), often derived from problem-solving experiences which support their degree of relevancy (Castela, 2009).

The legitimating process of a technique generates, for each standard in the institution, elements of discourse that belong to the technology and are acknowledged by the institution as accepted assertions in relation to the technique. As well as the technique, the logos component of a praxeology is institutionally marked:

> The notions of technology and theory must be understood in a way that is spe-
> cific to the institution or person concerned. Technological is what, in an institu-
> tion or for a person, fulfils the technological function [...] Similarly, theory is
> what assumes, in that institution or for that person, a theoretical function.
>
> *(Chevallard, 2007, p. 714)*

Chevallard's comment helps us to see, first, that the process and criterions for assertions acknowledgement depend on the institution; they may be based, as seen above, on discursive theoretical proofs as well as on experimental or even empirical processes. Second, that for the ATD, a theory is not necessarily what science considers a theory to be—that is, an organised corpus of concepts, proofs and theorems. For instance, in indigenous contexts, the assertion "We use this technique because our parents have taught us to do so", recurring for the familial practices, will be considered as providing the theory of the familial praxeological organisation (for a discussion about theory in the ATD, see Castela & Elguero, 2013; Chevallard, Bosch & Kim, 2015).

To sum up the above considerations, we use the following diagram [T, τ, θ, Θ] ← I to indicate the processes of creation, legitimation and institutionalisation of a praxeology [T, τ, θ, Θ] within and by an institution I.

7 Refining the technological analysis to identify institutional differences

A direct consequence of the previous considerations regarding the idiosyncratic nature of praxeologies is the following ATD hypothesis: When a praxeology, produced within a given institution I, moves to another institution I_u to be used, this will most likely result in some transformations, called transpositive effects. In the transposition process, I_u adapts the imported praxeology to its own ecology, resources, needs and values, which can lead to changes in all elements of the praxeology. Any didactic institution I_d that intends to train students to meet I_u's demands should be aware of these changes from I to I_u; otherwise, they will leave the full responsibility of the praxeological adaptation up to the students. Moreover, let us recall that the specific nature of activities within I_d, under specific constraints, results in other transpositive effects, the so-called didactic transposition (Chevallard, 1985/1991). We discuss the issue of transposition in the next section. In this section, we want to present the kind of praxeological, especially technological, analysis developed by Corine Castela and Avenilde Romo (2011) in a case where several institutions are implied.

7.1 An example related to engineering sciences

The type of tasks on which we focus here is the following one: "Breaking up a rational function into partial fractions"—which we note as *BUP*. We can meet *BUP* in strictly mathematical contexts as well as in engineering sciences.

7.1.1 The mathematical praxeologies for BUP

Within the limits of this contribution, we only give a few elements about the mathematical praxeologies (for more details, see Castela, 2016, 2017). Let us note that, except for examples of effective calculation, everything below belongs to the logos block of the praxeology.

Description of the technique in the general case[1]:

1. Make the denominator monic (leading coefficient 1), and use the Euclidean algorithm to reduce to a problem where the degree of the numerator r is less than the degree of the denominator d.
2. Factorise the denominator as a product of powers of distinct monic irreducible polynomials.
3. Write the fraction as a sum of partial fractions of the form R/Q^k where Q is one of the irreducible factors, k is at most equal to the multiplicity of Q in d and the degree of R is less than the degree of Q.
4. First, every R is unknown. The coefficients need to be determined. One way of doing this is to take a common denominator, multiply out, equate coefficients and solve the resultant system of linear equations.

Example. Let us apply the technique to

$$\frac{3x + 1}{(x - 1)^2(x + 2)}.$$

We obtain:

$$\frac{3x + 1}{(x - 1)^2(x + 2)} = \frac{A}{(x - 1)} + \frac{B}{(x - 1)^2} + \frac{C}{x + 2}$$

$$\Leftrightarrow 3x + 1 = A(x - 1)(x + 2) + B(x + 2) + C(x - 1)^2$$

$$\Leftrightarrow 3x + 1 = (A + C)x^2 + (A + B - 2C)x - 2A + 2B + C$$

$$\Leftrightarrow \begin{cases} A + C = 0 \\ A + B - 2C = 3 \\ -2A + 2B + C = 1 \end{cases} \Leftrightarrow \begin{cases} A = 5/9 \\ B = 4/3 \\ C = -5/9 \end{cases}$$

This technique [τ_1] is heavy in some cases, with many coefficients to find, if you do not have any software to solve the system. Indeed, mathematicians look for other techniques. For example, in the following excerpt from an online calculus textbook[2], we find an appraisal of τ_1 when the denominator is a product of n linear terms followed by the presentation of another technique τ_2:

> However, for $n \geq 3$ the procedure becomes messy because we first need to do a lot of tedious term multiplication to find coefficients, and then we need to solve a tedious system of linear equations. [...] The approach we will now use is based on the key idea that if two polynomials are equal, their values at every number are equal.
>
> *(Ibid., p. 7)*

Using τ_2 on the previous example gives the following:

$$3x + 1 = A(x - 1)(x + 2) + B(x + 2) + C(x - 1)^2$$

so taking $x = 1$ and $x = -2$, we get every term but one equal to 0, so that we have $4 = 3B$ and $-5 = 9C$. To get A, there is no special value that would eliminate B and C. We can choose $x = 0$ and get $A = 5/9$. According to the online textbook, this technique is clearly "preferable (from a speed perspective) when $n \geq 3$" *in the case of linear factors.*

We can ask why it is important to make the denominator monic. In the textbook above we find no explicit motivation for this choice. However, the fact that linear monic polynomials have 1 as a derivative and hence that the antiderivative of rational functions $1/(x - a)^k$ is easy to calculate, can be one motive for the restriction to monic factors.

7.1.2 The automatics praxeology for BUP

The example we consider now is from an online course in automatics for higher technicians[3] (Romo, 2009). One of the objectives of automatic control is to "govern" the behaviour of the systems. The signal to be controlled is compared to the desired reference, the discrepancy is then used to compute corrective actions. The less time to get the signal back to the expected reference, the more efficient is the control system.

The temporal evolutions of the different systems involved are described by differential equations, turned to algebraic ones by the Laplace transform[4] and relatively easily solved, with a rational fraction $F(p)$, called the *transfer function* of the system, as a solution. In the end, you have to get back to the temporal function—that is, to inverse the Laplace transform. The online textbook recommends using a table of Laplace transforms, specially adapted to automatics requirements. The type of tasks *BUP* appears when complicated rational fractions $F(p)$ are involved. What follows gives an idea of the technique and technology proposed by the textbook.

> *Description of the technique*: The author assumes that the mathematical techniques we have presented before are familiar to the students. He specifies only that $F(p)$ denominator must be written under the following canonical form $k(1 + \tau_1 p)(1 + \tau_2 p)(1 + \tau_3 p)\ldots$ with decreasing values of the τ_i. For instance, $3p + 2$ is transformed into $2(1 + 1.5p)$ and not into $3(p + 2/3)$. This is a significant change to the mathematical technique.

The motivation for this special factorisation is based on the following fact. If $F(p) = \frac{1}{1+1.5p}$, the corresponding original function is $f(t) = K(1-e^{-t/1,5})$. Value 1.5 is called the *time constant* of this function. The system reactivity, hence its quality, is directly dependent on the time constants τ_i, more precisely on the higher value; therefore, this value must appear clearly in the calculation. Moreover, if $f(t)$ represents the controlled quantity and K its desired constant value, it is known that after 7τ, here 7×1.5 seconds, the exponential will be equal to 0, that is, considered as negligible in automatics. Hence the transitional regime lasts 7×1.5 seconds. The validation is relatively straightforward: $e^{-t/\tau} < 0.001 \Leftrightarrow e^{t/\tau} > 1000 \Leftrightarrow t > \tau\ln(1000) \approx 7\tau$.

7.2 What needs does the technology of a technique intend to satisfy?

The previous example is from an investigation that addresses the issue of the mathematical preparation of engineers (Romo, 2009). Due to the central role played by the Laplace transform in engineering, it includes a comparative study on the way this notion is taught in the textbooks of different university institutions, one written by a mathematics lecturer, the other two by lecturers in automatics. The first one focuses on the comprehensive, accurate presentation of concepts, theorems and proofs. Automatics textbooks are very different. They give a lower priority to mathematical proofs and instead develop another kind of knowledge about techniques, strongly correlated with the vocational context. There are many things to know about the Laplace transform and derived techniques, but all these technological elements satisfy diverse needs.

Drawing on the aforementioned textbooks, Castela and Romo (2011) have differentiated six roles of the technology we have illustrated in the previous examples: (1) describing the technique; (2) validating it (i.e. proving that this technique produces what is expected from it); (3) explaining the reasons why this technique is effective (knowing about causes); (4) motivating the different steps of the technique (knowing about objectives); (5) making it easier to use; and (6) appraising it (with regard to the field of effectiveness, to the using comfort, relative to other available techniques, etc.). This list should not be taken as exhaustive. For instance, drawing on other investigations (e.g. Covián Chávez, 2013, on land surveyor training), one more role is currently considered: controlling the technique implementation. Even if a mathematical proof justifies that the technique, when adequately implemented, produces the expected solution, an individual may make errors when using it. The institution where this individual works requires that she or some supervisor be able to verify the process.

Coherently with our hypotheses about the relations between the technique standards and the technology in a given institution, we can associate these technological categories with the categories of technical standards evoked in section 5. Describing the technique, which supposes the elaboration of ostensive objects (words, schemes, etc.), contributes to the objectification of individual ways of doing, hence to the interactions between the different actors within the institution, to the dissemination of the technique and to any other processes of legitimation which need the technique to be a social object. Validating the technique as well as appraising its scope relates to what we have called validity standards, appraising its using comfort and increasing its usability, controlling the implementation for safety and reliability. Lastly, motivating and explaining are two aspects of intelligibility: what is the rationale for doing that? Why does doing that lead to the desired effect?

In short, this refined kind of technological analysis is a tool to investigate the institutional influence on praxeologies and therefore the transposition phenomena.

8 When praxeologies move from one institution to another: the transpositive effects

Avenilde Romo (2009) considers four institutions: two scientific research institutions, in mathematics $[I_r(M)]$ and in automatic control $[I_r(AC)]$, and the related university institutions in different vocational courses $[I_u(M), I_u(AC)]$. The social responsibility of the first two is to produce new praxeologies in their field. In the case of the BUP praxeology we have considered in section 6 and more generally for the differential equations solving praxeologies related to the Laplace transform, $I_r(M)$ nowadays acts as a praxeology producer. $I_r(AC)$ uses $I_r(M)$'s praxeologies, but the movement from the mathematics institution to the automatic control one changes the praxeologies. Then the movement goes on from $I_r(AC)$ to $I_u(AC)$, with new possible transpositive effects because students are beginners, and to the working context perspective. We have seen that in the BUP praxeology, the type of tasks and the technique have changed due to the requirement about the time constants. If we consider the technology, we may assume that there is no change regarding the theoretical validation of the technique itself, mainly based on mathematical properties, but we find new elements coming from the fact that the mathematical type of tasks has been embedded in an automatic control type of tasks and is submitted to different technical standards.

The diagram in Figure 4.3 gives a general representation of the possible transpositive effects when a praxeology $[T, \tau, \theta r, \Theta]$ produced by a research institution I_r moves to be used in an institution I_p which has a pragmatic relation to T and the correlated praxeology. I_p may be an educational institution working with the praxeology to teach it.

Let us consider the different symbols in this figure which generalises the mathematics case.

8.1 The original praxeology

I_r is a research institution, namely an institution socially in charge of producing new praxeologies to address certain types of tasks and organising systematic processes of validation, to substantiate their legitimacy and institutionalisation. It should be underlined that the validation processes depend on the specific paradigm of I_r. Besides, θ^r may hold other elements related to the I_r technique standards. For instance, in mathematics, maximal effectiveness is looked for through theorems with minimal hypotheses. This is not necessarily the case in other sciences where the models try to use objects with the least possible sophistication (e.g. regular functions).

8.2 The transposed praxeology

What does this diagram say? At first, the asterisk expresses that every component of the original praxeology may evolve. This transformation is an object of institutional transactions completed in a specific institution I^*_r, created and controlled by I_r and I_p, an institution in charge of the transposition, that Chevallard (1985/1991) calls a *noosphere*. I^*_r is more or less vanishing; the transactions are more or less difficult and controversial, depending on several factors. One of these factors is the extent of the transformations, the distance between the two institutional epistemologies (e.g. I_r is

$$[T, \tau, \theta^r, \Theta] \leftarrow I_r \curvearrowright \left[T^*, \tau^*, \begin{matrix} \theta^{*r}, & \Theta^* \\ \theta^p, & \Theta^p \end{matrix}\right] \begin{matrix} \leftarrow I^*_r \lessgtr^{I_r}_{I_p} \\ \leftarrow I_p \end{matrix}$$

FIGURE 4.3 From I_r to I_p, the transpositive effects model

mathematics and I_p is an experimental science). Another factor is the importance for I_p that I_r validates the new technique (e.g. I_p is a medical profession with high-security requirements) or the importance for I_r that the transposed praxeology be not too far from the original one (frequently an issue when I_p is a secondary education institution). Finally, this diagram says that a practical technology θ_p is developed and acknowledged by I_p on specific empirical bases, sustained by a second-level discourse: a theory.

Romo (2009) shows that the transactions between institutions about the logos block $[\theta_r, \Theta]$ of the Laplace transform praxeologies may result in different forms of transposition depending on the vocational training institution considered. The crucial issue that underpins the transpositive choices is whether or not engineers or higher technicians should know about the mathematical validation and explanation of the techniques that they will be expected to use. In other words, whether—and to what extent—the training institution tries to reduce the presence of black boxes. Depending on the choice made, these praxeologies are taught by a mathematics lecturer or directly included in an engineering science course.

Romo (2009) meets the first case in a high-level engineering school. As mentioned, the $I_u(M)$ textbook is focused on the mathematical theory, exhaustively presenting proofs of the most extended theorems, working on sophisticated examples and counterexamples. In this case, the transposition remains totally under the control of mathematics; changes in $[\theta_r, \Theta]$ are limited. The two $I_u(AC)$ textbooks provide examples of the second case; they come from different vocational training institutions with an emphasis on occupational competencies. The analysis (Castela & Romo, 2011) reveals that the $[\theta_r, \Theta]$ block partially vanishes: the mathematical proofs of some theorems (mainly the easiest ones, with low theoretical needs) are presented; for other theorems, the proof is omitted, but the textbook refers to its existence as a mathematical guarantee; finally, the problematic dimension of some assertions is completely hidden. The counterpart to these choices relative to the mathematical component of the technologies is that in the $I_u(M)$ textbook, the practical component is missing, while it is specifically developed in the $I_u(AC)$ textbooks, with an emphasis on the technological elements satisfying motivation and appraisal needs.

8.3 Other dynamics

So far, we have only considered a specific case—namely, a praxeology produced by a research institution moving to another institution to be used or taught. Yet we cannot assume that the praxeological production is exclusively developed in research institutions. Occupational and, more broadly, social settings may develop their own original praxeologies, $[T, \tau, \theta^p, \Theta^p]$, within the empirical contexts of working and social life. These praxeologies move to other institutions to be used with possible transpositive effects. Another possibility is that a research institution, created or not by the occupational one, investigates the original praxeology to improve it and more systematically substantiate its validity. As an example, Corine Castela and Cecilia Elguero (2013) examine the case of custom dressmaking in Argentina, with various sizes of institution involved.

This approach is an incentive for vocational institutions to analyse the nature and extent of the transpositive effects on mathematics praxeologies within the scientific and professional fields included in their curriculum. As seen in our examples, each component of the praxeology may change or develop for rational reasons that take into account the specific conditions of activities. Educational institutions should consider the motives and legitimacy of these evolutions. More generally, this style of analysis, which pays close attention to the institutional specificities of praxeologies, should protect us against naïve approaches of interdisciplinarity at school that usually underestimate the distance between disciplines and fail to provide teachers with tools to design authentic interdisciplinary sequences.

Notes

1 It should be noted that the technique is described as a set of types of tasks.
2 Partial fractions: an integrationist perspective. Math 153 Section 55. Vipul Naik. University of Chicago. https://vipulnaik.com/math-153-sequence.
3 http://public.iutenligne.net/automatique-et-automatismes-industriels/verbeken.
4 The Laplace transform is a linear integral operator that transforms $f(t)$ with a real argument t $(t \geq 0)$ to a function $F(p)$ with complex argument p. It transforms $f'(t)$ to $p \cdot F(p)$.

References

Bosch, M., & Chevallard, Y. (1999). La sensibilité de l'activité mathématique aux ostensifs. *Recherche en Didactique des Mathématiques*, 19(1), 77–124.

Brousseau, G. (1981). Problèmes de didactique des décimaux. *Recherches en Didactique des Mathématiques*, 2(1), 37–127.

Brousseau, G. (1994). *Problèmes et résultats de didactique des mathématiques*. Washington: ICMI Study.

Brousseau, G. (2002). *Theory of didactical situations in mathematics*. Dordrecht, The Netherlands: Springer Netherlands.

Castela, C. (2009). An anthropological approach to a transitional issue: Analysis of the autonomy required from mathematics students in the French Lycée. *NOMAD*, 14(2), 5–27.

Castela, C. (2016). Cuando las praxeologías viajan de una institución a otra: Una aproximación epistemológica del "boundary crossing". *Educación Matemática*, 28(2), 8–29.

Castela, C. (2017). When praxeologies move from an institution to another one: An epistemological approach of boundary crossing. In R. Göller, R. Biehler, R. Hochmuth & H.-G. Rück (eds), *Didactics of mathematics in higher education as a scientific discipline: Conference proceedings* (pp. 418–425). Kassel, Germany: University of Kassel.

Castela, C., & Romo, A. (2011). Des mathématiques à l'automatique : Étude des effets de transposition sur la transformée de Laplace dans la formation des ingénieurs. *Recherches en Didactique des Mathématiques*, 31(1), 79–130.

Castela, C., & Elguero, C. (2013). Praxéologie et institution, concepts clés pour l'anthropologie épistémologique et la socioépistémologie. *Recherches en Didactique des Mathématiques*, 33(2), 79–130.

Chaachoua, H. (2010). *La praxéologie comme modèle didactique pour la problématique EIAH. Etude de cas: La modélisation des connaissances des élèves* (Note de synthèse HDR). Grenoble, France: Université Joseph Fourier.

Chevallard, Y. (1985/1991). *La transposition didactique. Du savoir savant au savoir enseigné.* Grenoble, France: La Pensée Sauvage (2nd edition).

Chevallard, Y. (1999). L'analyse des pratiques enseignantes en théorie anthropologique du didactique. *Recherches en Didactique de Mathématiques*, 19(2), 221–265.

Chevallard, Y. (2006). Steps towards a new epistemology in mathematics education. In M. Bosch (ed.), *Proceedings of the IV Congress of the European Society for Research in Mathematics Education* (pp. 21–30). Barcelona, Spain: FUNDEMI-IQS.

Chevallard, Y. (2007). Passé et présent de la théorie anthropologique du didactique. In L. Ruiz-Higueras, A. Estepa, & F. J. García (eds), *Sociedad, escuela y matemáticas: Aportaciones de la teoría antropológica de lo didáctico* (pp. 705–746). Jaén, Spain: Publicaciones de la Universidad de Jaén.

Chevallard, Y., Bosch, M., & Kim, S. (2015). What is a theory according to the anthropological theory of the didactic? In K. Krainer & N. Vondrová (eds), *Proceedings of the 9th Congress of the European Society for Research in Mathematics Education* (pp. 2614–2620). Prague: University of Prague.

Cirade, G., & Matheron, Y. (1998). Équation du premier degré et modélisation algébrique. In R. Noirefalise (ed.), *Actes de l'École d'Été de la Rochelle* (pp. 199–250). Clermont-Ferrand, France: IREM de Clermont-Ferrand.

Covián Chávez, O. N. (2013). La formación matemática de futuros profesionales técnicos en construcción (Doctoral dissertation). CINVESTAV-IPN, Mexico.

Croset, M.-C. (2009). Modélisation des connaissances des élèves au sein d'un logiciel éducatif d'algèbre : Étude des erreurs stables inter-élèves et intra-élève en termes de praxis-en-acte (Doctoral dissertation). Université Joseph Fourier, Grenoble, France.

Croset, M.-C., & Chaachoua, H. (2016). Une réponse à la prise en compte de l'apprenant dans la TAD: La praxéologie personnelle. *Recherches en Didactique des Mathématiques*, 36(2), 161–196.

Pilet, J. (2012). Parcours d'enseignement différencié appuyés sur un diagnostic en algèbre élémentaire à la fin de la scolarité obligatoire: Modélisation, implémentation dans une plateforme en ligne et évalution (Doctoral dissertation). Université Paris Diderot, France.

Roditi, E. (2014). Le calcul des doses médicamenteuses. Pratiques professionnelles et choix de formation en soins infirmiers. *Recherches en Didactique des Mathématiques*, 34(2/3), 103–132.

Romo, A. (2009). La formation mathématique des ingénieurs. (Doctoral dissertation). Université Paris Diderot, France.

Tang, M. D. (2014). Une étude didactique des praxéologies de la représentation en perspective dans la géométrie de l'espace, en France et au Viêt Nam (Doctoral dissertation). Université Joseph Fourier, Grenoble, France.

5

RESEARCH ON NEGATIVE NUMBERS IN SCHOOL ALGEBRA

Eva Cid, José M. Muñoz-Escolano and Noemí Ruiz-Munzón

1 A reference epistemological model on elementary algebra

Several works in the field of the ATD have questioned the conception of elementary algebra prevailing in school institutions (Bolea, Bosch & Gascón, 2001; Bosch, 2015; Chevallard, 1989, 1990; Gascón, 1993, 1994/5). This conception, usually called *generalised arithmetic*, is characterised for considering algebra as a simple continuation of arithmetic. Algebraic expressions arise from the need of representing and handling unknown numbers, while algebraic tasks are reduced to the formal manipulations of algebraic expressions and the resolution of equations and inequalities. In both types of tasks, a distinction between known data and unknowns is established.

The questioning of this school conception has favoured a research line that looks for a school introduction of algebra understood as a functional tool to develop mathematical modelling activities (Bosch, 2015; Chevallard, 1989; Gascón, Bosch, & Ruiz-Munzón, 2017; Ruiz-Munzón, Bosch & Gascón, 2013, 2015). Due to the algebraised character of much of high school mathematics, we propose an introduction to algebra as a generic tool to model other mathematical school organisation—like operations with numbers, geometry, measure of quantities, etc.—giving rise to what has been called *an algebraisation process of a mathematical organisation* (Ruiz-Munzón, 2010). This process overcomes the school conception of algebra and broads its scope to link the (traditionally conceived as) algebraic work with the use of elementary functions as modelling tools. It corresponds to a reference epistemological model (REM, see chapter 6) built to both analyse the current school teaching of algebra and propose alternative instructional activities. The study of the institutional conditions that enable these proposals and of the constraints that hinder their development is still an open problem for research in didactics.

We start by briefly presenting the REM of the algebraic-functional modelling, where algebra appears as a tool to approach theoretical questions that arise in different areas of elementary mathematics, to generalise the resolution of tasks or problems appearing in these areas, and to structure them into types of tasks. We will then show how such a REM has evolved to integrate the introduction of negative numbers as part of the algebraic work. After presenting the main elements of the expanded REM, we will illustrate its use as a didactic tool to design new teaching sequences related to the introduction of integer numbers, following the work of Cid (2015).

1.1 First stage of the algebraic-functional modelling

The starting point of algebraic-functional modelling is the mathematical praxeological organisation A around "arithmetic problems", understood as problems that can be solved using a chain of arithmetic operations executable using the data of the problem. The chain of operations is called a *calculation programme* (CP) (Chevallard, 2005). We will note them by the expression $CP(a_1, a_2, ..., a_n)$ where $a_1, a_2, ..., a_n$ are the quantities to operate (or the *arguments* of the CP). We can consider that S covers most of the elementary arithmetic that is traditionally taught at primary school. The techniques of S consist of the execution of CPs using verbal discourses accompanying written operations. Its technological-theoretical elements are those of the arithmetical operations that build and justify the chain of operations to calculate the unknown, including some basic properties of quantities, measures, numbers and operations.

A first algebraic modelling of the previous mathematical organisation could be obtained by substituting the CP expressed rhetorically by a written expression, which gives the CP certain materiality. This is useful when one has to consider a CP as a whole, not as a process decomposed into successive steps. This passage, from a mainly oral work to a written one, gives rise to a new mathematical organisation M_1. The kind of manipulations made with these written expressions is, essentially, their production, simplification (to obtain a CP equivalent to another one) and a comparison of their structures using equalities or inequalities between CPs. For instance, let us consider the following task:

> Think of a positive number, add its consecutive and add this consecutive number again, add 15 to the result and, finally, subtract three times the initial number. What is the result? Repeat the process with a different number. How is the final result modified? Can you explain this?

The first two questions can be answered in A, executing the described CP in a rhetorical form. If, for instance, the initial number is 4, we add the double of 5, that is 10, and obtain 14. We then add 15, getting 29, and subtract three times 4, that is, 12, to finally obtain 17. When the process is repeated with another number, the result is again 17. The search for an explanation for this phenomenon could be difficult if we remain in the rhetorical work. It becomes easier when we pass to an algebraic modelling of the CP in which the chosen number is substituted by a letter, n, or any other symbol. In this way, the CP can be written as $n + 2(n + 1) + 15 - 3n$ and, transforming this expression conveniently, it leads to:

$$n + 2(n + 1) + 15 - 3n = n + 2n + 2 + 15 - 3n = 3n + 17 - 3n = 17.$$

This explains the result obtained and even ensures that it will be the same as with any other initial number. We can see that M_1 has to include techniques to produce a CP equivalent to a given one and the theoretical environment necessary to describe and justify these techniques. We have used the sign "=" here as an abbreviation of the expression "leads to". It could have been replaced by an arrow:

$$n + 2(n + 1) + 15 - 3n \rightarrow n + 2n + 2 + 15 - 3n \rightarrow 3n + 17 - 3n \rightarrow 17.$$

1.2 Second and third stages of the algebraic-functional modelling

The second stage of the REM arises when one wants to approach questions that imply the joint manipulation of CPs to answer in terms of relations between the arguments of the CPs. This leads to a new mathematical organisation, M_2, which is an extension of M_1. M_2 is characterised by the type of tasks solved using equalities between CPs. The theoretical block of M_1 is then considerably expanded to include a new interpretation of the sign "=" as an indicator of an equivalence (as in $8 + n - 5 = n + 3$), and not only as the result of an arithmetical operation (as in $4 + 2 = 6$). It also has to include the development of such equation techniques as, for instance, the cancellation of terms. In summary, M_2 includes the praxeologies of the equational calculus and the consideration of the equations as new mathematical objects.

Let us see an example of a task placed totally in this second stage:

> Eva and Bernardo play a card game in which they bet counters of two colours (red and white). Counters of the same colour have the same number of points. At the end of the game, Eva has 20 white counters and 90 red ones, and Bernardo has 40 white counters and 60 red ones. How many points correspond to the white and red counters if, at the end, Eva and Bernardo have the same number of points?

If w denotes the number of points of a white counter and r the number of points of a red counter, the CPs "giving us the number of points of Eva and Bernardo" are, respectively, $20w + 90r$ and $40w + 60r$. To answer the first question, it is not enough to simplify the CP separately; it is necessary to establish the equality between them and apply cancellation techniques: $20w + 90r = 40w + 60r$. Therefore $30r = 20w$ and $3r = 2w$.

The third stage of the algebraisation process is an extension of M_2. It arises with the incorporation of questions related to the joint variation of two or more arguments and its impact on the CPs' variation. The mathematical organisation M_3, which contains M_2, includes techniques based on the production and manipulation of algebraic *formulas*, without necessarily distinguishing unknowns and parameters. From M_2 and M_3, new extensions are possible. For example, the interpretation of formulae and equations as functional models leads to praxeologies including the techniques of differential calculus (chapter 6). Finally, in Ruiz-Munzón (2010), the REM presented is completed with the design of instructional processes for the introduction of algebra in grade 7 (12–13 years old) and the transition from algebra to functional modelling in grades 10 and 11 (15–17 years old).

2 A reference epistemological model on the introduction of negative numbers

Nowadays, and especially after the New Maths reform, the introduction of negative numbers in schools is made in an arithmetic environment and relies on specific models based on the presentation of opposite or relative quantities. Research has questioned the relevance of such an introduction. In some cases it is shown that the analogy with specific models could interfere with a correct building of the algebraic structure of signed numbers (Gallardo, 1994). In other cases, epistemological studies show that the raison d'être of signed numbers is the algebraic work, which explains why they were so difficult to be accepted historically (Brousseau, 1983; Cid, 2015; Glaeser, 1981; Schubring, 1986).

If we consider negative numbers as part of the algebraic work, we need to include them in the REM about algebraic modelling we have just presented. Two important points need to be noticed here. On the one hand, the introduction of algebra should be done at the same time as the introduction of negative numbers, because algebraic manipulations cannot be done without the "rule of signs" (plus by plus is plus, plus by minus is minus and minus by minus is plus). Therefore, the presentation of new algebraic objects and of techniques to manipulate them is conditioned by the need to introduce negative numbers. On the other hand, even if elementary algebraic work can be done with rational numbers, it can be important to separate the additive symmetrisation of N towards Z from the one of Q^+ towards Q, leaving the later for a subsequent stage. According to this, the REM starts from the need to consider the *addends* and *subtrahends* as elements of the calculation programmes, and it finishes interpreting them as a new numerical set, which extends the positive integer numbers.

The initial system to be modelled is also the system of the arithmetic problems A but with two constraints. On the one hand, only additive CPs are considered—that is, CPs without products or quotients. Therefore, the set of problems is reduced to problems that can be solved through a sequence of additions and subtractions starting from the initial data. The common arithmetical techniques consist in making the operations one after the other, using the result of the previous one as data for the following one. On the other hand, to force the passage from A to M_1, in the type of arithmetical problems considered, some of the initial data required to obtain the solution are not given. Therefore, arithmetical techniques cannot be applied unless the unknown data is written or symbolised in some way.

We thus move from an arithmetical activity based on the effectuation of a sequence of simple CPs to an activity more typical of algebra: solving types of problems by obtaining one or more formulas that model them. This states better the epistemological rupture between algebra and arithmetic. It also allows a smoother entry into algebraic calculations because equation-solving techniques are postponed: only techniques of establishing and simplifying simple algebraic expressions are established. Therefore, all the stages of the REM of negative numbers can be considered as included in M_1. Although algebraic objects appear that could be considered belonging to M_2, and even to later stages, their manipulation techniques are not developed.

2.1 The first stage of the introduction of negative numbers

Let us call N_1 the subset of M_1 problems in which a first (cardinal) number suffers different increases or decreases that lead to a final cardinal that is the requested solution. For instance:

> A train leaves Barcelona with a certain number of passengers and arrives in Girona after making two stops. At the first stop, 15 passengers get off, and 12 passengers get on; at the second stop, 38 get off, and 42 passengers get on. With how many passengers did the train get to Girona?

The lack of the initial data or one of the intermediates—here, the initial number of passengers in Barcelona—prevents carrying out the CP necessary to obtain the solution. To "give a solution" to the problem, one has to use a letter or any other symbol to indicate the operations to perform. This leads to the introduction of additive algebraic expressions: the formula that solves the problem. For instance, in the previous problem, the existence of an unknown quantity transforms the arithmetic problem into an algebraic one. A CP to model the situation is $b - 15 + 12 - 38 + 42$, where b is the initial number of passengers in Barcelona. The need to simplify it in order to obtain a more efficient formula and the impossibility of carrying out the first operation makes it necessary to think in terms of "subtracting 15 and adding 12 to a number is the same as subtracting 3 from it", which transforms the CP into $b - 3 + 4$ and then into $b + 1$. Therefore, the simplification of algebraic expressions requires calculations and reasoning in terms of *addends* and *subtrahends*, which become new objects of the praxeology. The unknown quantity does not need to be at the beginning of the CP.

Let us see another example:

> Maria goes shopping. She has €120. She first buys a pair of trousers that costs €40 and later on a pair of shoes. Finally, she buys a book, which costs €15.

How much money does Maria have? If Maria has €30 left, can we know how much the shoes are?

A CP to model the situation is $120 - 40 - p - 15$, p being the price of the shoes. The first operation can be carried out as a subtraction between quantities, and we obtain $80 - p - 15$. Now, the manipulation of this last expression requires us consider the numbers or the letters together with the sign preceding them (addends and subtrahends). Therefore, as "subtracting first p and later 15" obtains the same result as "subtracting first 15 and later p", we can write $80 - p - 15 = 80 - 15 - p = 65 - p$. Again, the simplification of algebraic expressions requires calculations and reasoning in terms of addends and subtrahends.

In the second part of the task, we obtain an equation, $65 - p = 30$, which can be solved in M_1 without any equation-solving technique. However, the reinterpretation of sums and subtractions as the compositions of addends and subtrahends (the compositions of translations) allows for a greater degree of manipulation of algebraic expressions since this composition has properties (associative, commutative, existence of neutral and opposite elements) that contribute to a greater economy and justification of algebraic calculations. For instance, a CP like

$$45 - f + g - 10 + 532 + f - 532 + 19 + 27 - 9 - 21$$

can be rewritten, thanks to the communicative properties, into

$$45 - f + f + 532 - 532 + 19 - 9 - 10 + 27 - 21 + g.$$

The associative properties allow one to choose what parts of the CP to operate on first, leading to the equivalent CP, $51 + g$. Therefore, the first mathematical organisation moves from additions and subtractions between absolute numbers to the composition of addends and subtrahends. This also transforms the meaning of the signs "+" and "−" from being *binary operators* (signs that in arithmetic indicate a binary operation between absolute numbers) to being *generalised binary operators* (signs that in algebra indicate that the number that follows them has a role as addend or subtrahend). The elements constituting the technology of S are extended and, accordingly, the associated resolution techniques.

2.2 The second stage of the introduction of negative numbers

A second stage, N_2, is developed when priority is given to the *difference* as the result of the action of comparing, in contraposition to the *subtraction* as the result of the action of removing. The passage from N_1 to N_2 is motivated by questions on the comparison of quantities given by CPs. We continue imposing the restriction that some variables are unknown and that CPs can be reduced to a first-degree polynomial expression. This gives rise to a new problem: the need to make operations of operations—that is, the need to add or subtract terms that in turn are additions or differences that cannot be effected, which forces the use of parentheses in algebraic expressions and the establishment of equivalences between expressions with parentheses and without parentheses. In N_1, the sign "−" indicated the subtraction of a term from another. From now on it must also be interpreted as a sign that causes a change in the condition of addends or subtrahends, transforming an addend into a subtrahend and vice versa. Furthermore, this introduces a new object: a "signed difference". It can be interpreted as an absolute value quantifying the difference between two numbers and a sign indicating that the first term of the difference is greater or smaller than the second term.

Let us take an example:

Javier has a certain number of cards, Carmen has five more cards than Javier and Carlos has twice as many as Javier. If Javier and Carmen join their cards, will they have more or fewer cards than Carlos? Who has most cards, Javier, Carmen or Carlos? And who has fewest cards?

If we denote the number of cards that Javier has by c, then Carmen will have $c + 5$ cards and Carlos $2c$ cards. If Javier and Carmen join their cards, they will together have $2c + 5$ cards. To answer the posed questions it will be necessary to compare the expressions $2c + 5$ and $2c$, c and $c + 5$, c and $2c$, and finally $c + 5$ and $2c$, where c can only take positive integer values. The three first comparisons can be solved applying simple arithmetical techniques (from N_1), concluding that inequalities $2c + 5 > 2c$, $c < 2c$ and $c < c + 5$, are satisfied for all values of c. However, the comparison between $2c$ and $c − 5$ depends on the value of c. The difference $2c − (c + 5) = c − 5$ implies that $2c$ will be greater, equal or less than $c + 5$ depending on c being greater, equal or less than 5. Here we have another example of an algebraic object, an inequation, which will not be systematically treated in N_2.

The need to quantify the differences between additive algebraic expressions leads to a new extension of the technological elements of N_1. The sign "−" is reinterpreted as a *unary operator* sign (a sign transforming an addend into a subtrahend and vice versa). This allows establishing the equivalence between expressions with brackets and without brackets, which leads to the rules of brackets elimination.

2.3 The third stage of the introduction REM of the negative numbers

The third stage of the REM can be represented by a mathematical organisation N_3 that includes the product of CPs from N_2. We can think, for instance, in a praxeology constituted by geometric problems modelled arithmetically, to calculate the area of flat figures with some unknown data or to compare areas of figures when the sides are increased or decreased by a given quantity. Here the arguments of the CPs represent measures of length, although their domain remains that of natural numbers.

When any of the sides of the figure is not known, the area must be calculated by multiplying a natural number by an additive algebraic expression and the comparison forces to manage parentheses multiplied by an addend or a subtrahend. As a consequence, the rules of signs corresponding to the product of addends and subtrahends are required and the importance of the distributive property in the algebraic calculation is shown, both to develop and to factorise expressions. Moreover, the technique of passing from the rhetorical formulation of an additive-multiplicative CP to its algebraic expression, or vice versa, requires assuming that the transformation of a quotient between two terms is the product of one of them by the inverse of the other. Let us illustrate this with some examples.

> We will start from the area of a rectangle given by $A = ba$, where b is the length of one side (the base) and a the length of the other side (the height). If we are told that $a = 3$ cm and that the base b increases by 2 cm, how much will its area have increased?

The expression of the area of the rectangle will be $A = 3b$ in the first case and $A = 3(b + 2)$ in the second case. The difference of areas is then $3(b + 2) - 3b = 3b + 6 - 3b = 6$ cm^2. Here appears the product of a natural number by an indicated operation and the need to use parentheses and to use the distributive property to be able to simplify the expression and obtain the difference.

> From a rectangle with a side of length 5 cm, we build two other rectangles: one in which we increase the known side by 3 cm and decrease the unknown side by 3 cm, and another in which we decrease the known side by 3 cm and increase the unknown side by 3 cm. What is the difference between the areas of the last two rectangles? What is the length of the unknown side if the two areas equal?

The expression of the areas of both rectangles is $A_1 = 8(a - 3)$ and $A_2 = 2(a + 3)$, where a is the value of the unknown side. We need to find its difference $A_2 - A_1 = 8(a - 3) - 2(a + 3)$, which includes the product of a subtrahend by an addition. Using the distributive property, we have $8(a - 3) - 2(a + 3) = 8a - 24 - (2a + 6) = 8a - 24 - 2a - 6 = 6a - 30$, the sign of which depends on the values of a. Both areas will be equal when $6a - 30 = 0$, that is when $a = 5$.

N_3 now includes additive-multiplicative PCs where it is necessary to manage the product of addends and subtrahends. The technological-theoretical elements have to include the corresponding sign rules. As we have seen, the distributive property becomes relevant to develop expressions as well as to extract common factors because it reduces the product between a number and a difference to a product between addends and subtrahends.

2.4 The fourth stage of the introduction of negative numbers

Finally, the activities corresponding to the fourth stage, which give rise to the mathematical organisation N_4, refer to the use of PCs where unknown arguments (letters) can take positive or negative values, independently of the preceding sign. This needs to give sense to the isolated addends and subtrahends and to introduce the *predicative* meaning of the signs "+" and "−" which indicate that the number that follows is positive or negative. The types of problems in N_4 include cases where unknown data, as well as the solutions, are vector quantities—that is, quantities which can be interpreted with a sense of either profit or loss, or one-way movement in a sense or the opposite, or as relative quantities. This corresponds to concrete models usually used at school for the introduction of integer numbers (lifts, wins and debts, over/below sea level, etc.), but the treatment is different.

The use of concrete models in the arithmetical domain does not seem enough to give integers a raison d'être. Instead, the use of these same models in an algebraic context, where the rules of calculation with addends and subtrahends are already established, allows showing the significant advantage of representing specific quantities using addends and subtrahends: the unification of the formulas. For example, we can calculate a difference of temperatures using a single formula, regardless of whether the temperatures are above or below zero. We can also calculate the distance between two cars starting from the same point regardless of which direction they move in because we can give the letters positive and negative numerical values. All this leads to the reinterpretation of addends and subtrahends as new numbers and the consolidation of a new meaning for signs "+" and "−": their predicative sense. They are no longer signs that indicate an operation, but signs that indicate a "quality" of numbers.

In the following example, we can see the technique used to unify the formula and work with the predicative meaning of the signs:

> Alberto plays cards. In the first game he loses three cards, in the second he does not remember what happened, in the third he wins five cards, and in the fourth he loses four cards. How many cards did he win or lose?

Now the nature of the task has changed. It is no longer a question of quantifying an initial or final number but the increase or decrease it has suffered. If we indicate by a the initial number of cards and by b the loss or gain of the second game, the total loss or gain will be given by

$$a - 3 + b + 5 - 4 = a - 2 + b, \; if \; b \; is \; a \; gain;$$

$$a - 3 - b + 5 - 4 = a - 2 - b, \; if \; b \; is \; a \; loss.$$

Here a and b are natural numbers. When giving values, we must use one of these two formulas depending on whether b is a loss or a gain. However, knowing the rules of calculation with addends and subtrahends allows us to assume that if b is replaced by the number preceded by the sign indicating whether it is a loss or a gain, the expression $-2 + b$ can be used in both cases. For instance, the unique formula $a - 2 + b$ can be used for b being a gain of 4 or a loss of 6:

$$a - 2 + b = a - 2 + (+4) = a - 2 + 4 = a + 2 \ and$$
$$a - 2 + b = a - 2 + (-6) = a - 2 - 4 = a - 6.$$

The unification of formulas provides a justification to denote with the signs "+" or "−" the measure of quantities with two directions or *relative quantities*. This allows defining or "giving sense" to isolated subtrahends as measures as well as oriented differences. It only remains to establish an order between addends and subtrahends to be able to consider them as numbers: the integers. And the order can be established from the necessity of having a total order compatible with the sum: a positive difference has to be compatible with a first term higher than the second; while a negative difference has to be compatible with a first term lower than the second. To sum up, the technological-theoretical elements of N_3 include the interpretation of "+" and "−" as predicative signs, the structure of integers as a totally ordered commutative ring with a unit element, and the representation of integer numbers in the real line to facilitate the ordering techniques.

3 Design of a teaching sequence

The REM about the introduction of negative numbers formulated in terms of a sequence of praxeological organisations $N_1 \rightarrow N_2 \rightarrow N_3$ was elaborated in interaction with the design and experimentation of a didactic sequence (Cid, 2015; Cid & Ruiz-Munzón, 2011). The sequence is structured in 53 tasks that, according to the theory of the study moments (see Glossary), respond to the following types:

- Problematic tasks (problems) that correspond to the *moment of the first encounter* and the *exploratory moment*. In these tasks, students meet a new problem and have to produce new techniques to solve it. The role of the teacher is to help the students in their deliberations, respecting their autonomy.
- Tasks that correspond to the *moment of the technical work:* individual tasks in which students are asked to exercise the previously established techniques to acquire greater control of them. The role of the teacher is to ensure the correct use of the techniques by helping students find appropriate elements of validation.
- Tasks that correspond to the *moment of the constitution of the technological-theoretical environment*, the *moment of institutionalisation* and the *moment of evaluation*. These are tasks shared by the teacher and the students, where students'

discoveries are shared and stated, transforming the knowledge of the teams of students into the knowledge of the class. Finally, the teacher will position this knowledge about the mathematical knowledge of society.

Many of the tasks, especially those relating to the technical work, are in turn broken down into several tasks. In tables 5.1, 5.2, 5.3 and 5.4, the stages of the didactic sequence are analysed from a praxeological point of view, detailing which algebraic objects emerge at each stage, which are institutionalised and which techniques of manipulation of these objects are developed systematically.

All the phases of the didactic sequence are considered included in stage M1 of the REM of algebraic-functional modelling. This stage corresponds to the explicit writing—using any written or graphical tool—of CPs and their simplification to obtain the required result. In this case, although the techniques that appear will be the germ of later techniques for solving equations and inequalities, no systematic work on these techniques is developed.

TABLE 5.1 Phase 1 of the didactic sequence

The first stage of the introduction of negative numbers (N_1)	
Type of tasks	Situations that can be modelled by additive CPs where an initial value or quantity is increased or decreased and one of the CP arguments is unknown. Solutions are additive CPs or their final result when the unknown is given a concrete value.
Algebraic objects that emerge (some of them remaining implicit)	– Generalised binary operative meaning of signs "+" and "–". – Operations as translations, composition of translations, properties, inverse. – Use of letters as variables, unknowns or parameters. – Change of variable. – Letter coefficients. – First-degree additive algebraic expressions, their sum and their numerical value. – Tables of values of related functions. – Equality between first-degree additive CPs in their dual aspect of identities and equations. – First degree equations with coefficients and solutions in N.
Techniques that are worked	– Symbolisation of CPs expressed verbally by first-degree additive algebraic expressions. - Simplification of first-degree additive CPs. - Composition of translations.
Technological and theoretical elements that are institutionalised	– Use of letters to indicate unknown numbers intervening as data in a problem. – Algebraic expressions as solutions to a problem. – Composition of translations and inverse translation. – Simplification of additive algebraic expressions without parenthesis, emphasising their non-algorithmic character.

TABLE 5.2 Phase 2 of the didactic sequence

The second stage of the introduction of negative numbers (N_2)	
Type of tasks	Situations that can be modelled by additive CPs where CP arguments are compared, and a question is asked about one of the arguments or the comparison when some of the data is unknown. Solutions are relationships of order between first-degree algebraic additive expressions or differences between them.
Algebraic objects that emerge (some of them remaining implicit)	– Unary operators; the meaning of the signs "+" and "–". – Difference between first-degree algebraic additive expressions and their symbolisation by parentheses. – Differences between addends and subtrahends. – Order relationship between first-degree additive algebraic expressions and their expression in terms of oriented difference. – Distinction between equalities/inequalities and equations/ inequations. – Solutions to first-degree inequations.
Techniques that are worked	– Calculation of the difference between two first-degree additive algebraic expressions. – Comparison of first-degree additive algebraic expressions by studying the sign of difference. – The suppression of parentheses and the simplification of first-degree additive algebraic expressions that contain them. – Symbolisation of arithmetic calculation programmes expressed verbally by first-degree additive algebraic expressions containing parentheses preceded by a "–" sign.
Technological and theoretical elements that are institutionalised	– Difference as a result of the action of comparing. – Reversibility of the order relationship. – Order between first-degree additive expressions as a function of the sign of their difference. – Brackets technique.

4 Implementation of a teaching sequence and conclusions

The implementation of the didactic sequence was carried out during the academic years 2008/9 and 2009/10 in a public school of Barcelona with two groups of 32 and 30 students in grade 7 (12–13 years, first year of secondary school in Spain) and the same groups of grade 8. In 2009/10 a revised proposal was implemented with a group of 22 students in grade 12. The school serves students aged 3–18 from families with medium to high cultural and economic levels. The students were participative and used to working in teams.

In grade 12, the tasks corresponding to the first three phases of the didactic sequence were implemented throughout 16 class sessions in 2008/9 and 17 class sessions in 2009/10. In grade 13, the tasks corresponding to the fourth phase were implemented, preceded by tasks to review the first phases. The experimentation was carried out in the schedule of mathematics classes: three hours a week. In the first two hours, students worked in small groups (four or five students per group), except for the wrap-up and institutionalisation moments. In the third session,

TABLE 5.3 Phase 3 of the didactic sequence

The third stage of the introduction of negative numbers (N₂)	
Type of tasks	Geometric problems such as the ones above that ask for the area of a rectangle when one of the sides is unknown or for the difference between areas when the sides are modified. Solutions are products of a natural number by an additive expression of first degree or differences between them. The domain of data and solutions remains that of natural numbers.
Algebraic objects that emerge (some of them remaining implicit)	– Product of a subtraction by an additive first-degree algebraic expression and its representation by parentheses. – Product of addends and subtrahends. – Multiplicative comparison between first-degree additive algebraic expressions.
Techniques that are worked	– Product of an addend or a subtrahend by a sum or a difference. – Use of distributive property to develop or factorise algebraic expressions. – Simplification of first-degree additive algebraic expressions with parentheses indicating products. – Symbolisation of arithmetic calculation programs expressed verbally by first-degree additive algebraic expressions containing parentheses indicating products.
Technological and theoretical elements that are institutionalised	– Product of an addend or a subtrahend by a sum or a difference. – Distributive property. – Hierarchy of operations in algebraic expressions. – Distinction between additive and multiplicative comparison.

students worked individually, exercising the techniques that had previously been used. Individual tests were also carried out to assess the students' learning.

The classes were given by the subject teacher with the assistance of an observer who recorded the events using personal notes, photos and photocopies of the students' work. The detailed diary of the class sessions can be found in Cid (2015). We summarise some of the conclusions of the experimentation as follows:

- The didactic sequence can be considered as viable, meaning that the construction of integers in an algebraic environment is feasible. Classes developed normally, and difficulty episodes could be solved quickly. Of course, the type of students, the educational institution, as well as the personal and professional characteristics of the teacher who took part in the experimentation, have turned out to be very favourable to the development of the didactic sequence. Predictably, in other educational environments, this proposal could encounter difficulties that have not been observed in this experimentation. In particular, the instructional sequence does not follow the current curricular organisation of contents, where integers are proposed to be learnt in grade 12 before the introduction of algebra in grade 13. The fact that the same teacher was responsible for both grades was judged to be an important element for this feasibility.

TABLE 5.4 Phase 4 of the didactic sequence

The fourth stage of the introduction of negative numbers (N₂)	
Type of tasks	Additive arithmetic problems where both known and unknown data and solutions can be interpreted as measures of quantities of two-way magnitudes or relative magnitudes. Initially, the domain of data and solutions remains that of natural numbers to end up being that of integers.
Algebraic objects that emerge (some of them remaining implicit)	– Predicative meaning of signs "+" and "–". – The fully ordered unitary commutative ring of integers. – Exact division between integers. – Differences between the algebraic structure of the set of integers and that of the set of natural numbers. – Numerical value of an algebraic expression when the variables domain is Z. – Order relation and difference between algebraic expressions when the variables domain is Z. – Complete notation that incorporates the binary operative signs between signed numbers.
Techniques that are worked	– Operations with integers. – Order with integers. – Obtaining the numerical value of a first-degree additive algebraic expression in Z. – Comparison of first-degree additive algebraic expressions in Z, studying the sign of difference. – Reducing complete notation to notation in which binary operative signs are suppressed (incomplete notation).
Technological and theoretical elements that are institutionalised	– Integers understood as natural numbers preceded by a "+" or "–" sign. – Equivalence between natural numbers and positive integers. – Opposite of an integer. – Absolute value of an integer. – Operations and order between integers. – Complete order of integers. – Properties of natural numbers and integers.

- We mention some of the constraints that seem to have hindered the development of the didactic sequence. The management of didactic time led to shortening students' processes of autonomous learning. The coordination of teamwork periods, the individual work and the pooling and institutionalisation moments was complicated due to the different rhythms of work of the groups. The students' difficulties with mental calculation techniques appeared to be a hindrance to the development of non-algorithmic algebraic calculations.

- Concerning the results obtained, we consider that the epistemological rupture sought, revealing the differences between the old (arithmetical) and new (algebraic) treatment of numbers did, indeed, occur. The difficulties students encountered in the resolution of tasks, the discussions that took place in groups and the initial resistance to abandoning the usual (arithmetical) way of calculating show the importance of the obstacle to be overcome.

- However, in the end, most of the students assumed the different meanings of the signs "+" and "−" and of the letters; they accepted the formula as the solution of a problem and the difference as the result of a comparison; and they internalised the principles of functionality and economy proper to algebraic calculus. The main pitfalls occurred in the interpretation of the sign "=" as a sign indicating a relationship and in the application of the technique of suppression of parentheses preceded by a sign "−".

- In the fourth stage, the experimentation showed that the didactic proposal had underestimated the epistemological leap from the generalised binary and unary operative meanings of the signs "+" and "−" to the predicative meaning. In spite of this, most of the students ended up going correctly from complete notation to incomplete notation in the additions and subtractions of whole numbers. By this we mean the passage of expressions like $a + (-3) − (+6) − (-4)$ to the simpler $a − 3 − 6 − (-4)$. However, fewer students correctly performed the products and ratios of whole numbers. As for the order between integers, its motivation by its compatibility with the addition (if $a < b$ and $c < d$ then $a + c < b + d$) did not produce any rejection in the classroom. In the end, it was mostly assumed that the sign of the difference indicates the order between the minuend and the subtrahend: if $a − b < 0$, then a < b; if $a − b > 0$, then $a > b$.

There exist different aspects that will be necessary to consider in order to continue completing and developing the REM of algebraic-functional modelling. One of these aspects is the extension of Z to the set of rational numbers, which implies a new development of M_1. Another aspect is the evolution of equational calculus, in what refers to the techniques of resolution of equations, as much as the techniques of operations between equations. This suggests an extension of the types of equations and systems to be studied in M_2 and M_3, the second and the third stages of the algebraisation process.

Finally, the lack of experimentation raises some questions about the compatibility between the REM on negative numbers ($N_1 \rightarrow N_2 \rightarrow N_3$) and the REM on the algebraic modelling process ($M_1 \rightarrow M_2 \rightarrow M_3$). For instance, in the first REM inequalities appear in N_1 while in the second REM they are only considered in M_2 or even M_3 … . Therefore, the possibility to extend M_2 and M_3 to situations that can be modelled with equations, linear systems of equations with more than two unknowns and inequations remains to be studied.

References

Bolea, P., Bosch, M., & Gascón, J. (2001). La transposición didáctica de organizaciones matemáticas en proceso de algebrización. *Recherches en Didactique des Mathématiques*, 21(3), 247–304.

Bosch, M. (2015). Doing research within the anthropological theory of the didactic: the case of school algebra. In S. J. Cho (ed.), *Selected regular lectures from the 12th International Congress on Mathematical Education* (pp. 51–69). Cham, Switzerland: Springer.

Brousseau, G. (1983). Les obstacles épistémologiques et les problèmes en mathématiques. *Recherches en Didactique des Mathématiques*, 4(2), 165–198.

Chevallard, Y. (1989). Le passage de l'arithmétique à l'algébrique dans l'enseignement des mathématiques au collège. Deuxième partie. Perspectives curriculaires: La notion de modélisation. *Petit x*, 19, 45–75.

Chevallard, Y. (1990). Le passage de l'arithmétique à l'algébrique dans l'enseignement des mathématiques au collège. Troisième partie. Perspectives curriculaires: Voies d'attaque et problèmes didactiques. *Petit x*, 23, 5–38.

Chevallard, Y. (2005). La place des mathématiques vivantes dans l'éducation secondaire: transposition didactique et nouvelle épistémologie scolaire. In C. Ducourtioux & P. L. Hennequin (eds), *La place des mathématiques vivantes dans l'enseignement secondaire* (pp. 239–263). Paris: Publications de l'APMEP.

Cid, E. (2015). Obstáculos epistemológicos en la enseñanza de los números negativos (Doctoral dissertation). Universidad de Zaragoza, Spain.

Cid, E., & Ruiz-Munzón, N. (2011). Actividades de estudio e investigación para introducir los números negativos en un entorno algebraico. In M. Bosch *et al.* (eds): *Un panorama de la TAD* (pp. 579–604). Barcelona, Spain: Centre de Recerca.

Gallardo, A. (1994). Negative numbers in algebra. The use of a teaching model. In J. P. da Ponte & J. F. Matos (eds), *Proceedings of the 18th International Conference of PME* (vol. 2, pp. 376–383). Lisbon: University of Lisbon.

Gascón, J. (1993). Desarrollo del conocimiento matemático y análisis didáctico: Del patrón análisis-síntesis a la génesis del lenguaje algebraico. *Recherches en Didactique des Mathématiques*, 13(3), 295–332.

Gascón, J. (1994/5). "Un nouveau modèle de l'algèbre élémentaire comme alternative à l'"arithmétique generalise". *Petit x*, 37, 43–63.

Gascón, J., Bosch, M., & Ruiz-Munzón, N. (2017). El problema del algebra elemental en la teoría antropológica de lo didáctico. In J. M. Muñoz-Escolano, A. Arnal-Bailera, P. Beltran-Pellicer, M. L. Callejo & J. Carrillo (eds), *Investigación en Educación Matemática XXI* (pp. 25–47). Zaragoza, Spain: SEIEM.

Glaeser, G. (1981). Epistémologie des nombres relatifs. *Recherches en Didactique des Mathématiques*, 2(3), 303–346.

Ruiz-Munzón, N. (2010). La introducción del álgebra elemental y su desarrollo hacia la modelización funcional (Doctoral dissertation). Universitat Autònoma de Barcelona, Spain.

Ruiz-Munzón, N., Bosch, M., & Gascón, J. (2013). Comparing approaches through a reference epistemological model: the case of school algebra. In B. Ubuz, Ç. Haser & M. A. Mariotti (eds), *Proceedings of the Eighth Congress of the European Society for Research in Mathematics Education* (pp. 2870–2879). Ankara, Turkey: Middle East Technical University.

Ruiz-Munzón, N., Bosch, M., & Gascón, J. (2015). El problema didáctico del álgebra elemental: Un análisis macro-ecológico desde la teoría antropológica de lo didáctico. *REDIMAT-Journal of Research in Mathematics Education*, 4(2), 106–131.

Schubring, G. (1986). Ruptures dans le statut mathématique des nombres négatifs. *Petit x*, 12, 5–32.

6

THE PHENOMENOTECHNICAL POTENTIAL OF REFERENCE EPISTEMOLOGICAL MODELS

The case of elementary differential calculus

Catarina Lucas, Cecilio Fonseca, Josep Gascón and Maggy Schneider

1 The notion of phenomenotechnique in didactics

The French epistemologist Gaston Bachelard (1949) conceives modern science as *phenomenotechnia*—that is, as the intelligent production of the phenomena researchers describe and explain. Here "phenomenon" should be understood in its scientific meaning, i.e. as an event or process that it is possible to typify and reproduce or, at least, that repeats over time. In order to clarify the potentially phenomenal nature of the sciences and, in particular, of didactic science, it would be necessary to analyse the process of construction of scientific phenomena in more detail, taking into account that this construction is carried out by scientific communities, by means of theoretical tools that are also used to formulate the associated research problems.

In the case of didactics of mathematics, the notion of didactic phenomena is not central in many approaches, although it played a crucial role in the birth of the theory of didactic situations (TDS, Brousseau, 2002). It is also essential in the approaches—such as the ATD—that follow the TDS in its ambition of constructing a didactic science. Many of the concepts proposed by these approaches, like "adidactic situation", "didactic contract", "didactic transposition", "praxeologies" or "study and research path", can be considered as theoretical tools to describe didactic phenomena. Moreover, the empirical confrontation of phenomena with the contingency makes it possible to raise new problems and continue progressing in the study of these phenomena. This highlights the phenomenological nature of didactic science.

In the following sections we will present some instruments didactics uses to "produce phenomena" and briefly describe the mechanism of this production.

1.1 Dominant epistemological models and reference epistemological models

The ATD postulates that, if an institution *I* carries out any kind of social manipulation of knowledge, whether its production, its teaching and learning or any type of application or use of it, then within the institution *I* a predominant way is developed (relatively shared and more or less explicit) of interpreting and describing the knowledge in question. This raises the notion of a *dominant epistemological model* (DEM) of a given piece of knowledge, or of a concrete domain, in *I*. In particular, we can talk about a DEM of each one of the mathematics' domains that are taught in a specific educational institution, as well as a DEM of the mathematics considered as a whole in this institution. The DEM of a domain of the mathematical activity assigns an *official raison d'être* to it—that is, a set of possible questions whose answer requires, in an essential way, the use of the knowledge components of that domain and, consequently, gives meaning to the school study of the domain. Both the DEM and the official raison d'être (of this domain in *I*) are usually quite transparent to and unquestioned by the subjects of *I*.

Based on this assumption, Gascón (2003) proposes to characterise the approaches and didactic theories that are part of the *epistemological program of didactics research* as those which question the dominant epistemological models of the educational institutions, explicitly propose *alternative epistemological models* of different domains and use them as a reference system to formulate and address didactic problems.

> Any didactic research that proposes to study phenomena related to a field of mathematics (e.g. elementary algebra), and in a given didactic institution, should not assume as such the implicit model prevailing in the institution, but should take it into account as an object of study, i.e. as part of the didactic facts that constitute the "empirical" basis of the research. To do this, the researcher needs a particular "point of view", i.e. an alternative model of the mathematical field of activity that serves as a reference framework for interpreting the dominant model in the institution under study.
>
> *(Gascón, 1994–1995, p. 44)*

Such provisional and relative models, built by the didacticians with heuristics and methodological purposes, are referred to as *reference epistemological models* (REMs, see Glossary). REMs facilitate a detachment from the DEMs of the considered educational institutions and help give visibility to phenomena that could otherwise remain unnoticed and unexplained. In particular, in the light of a REM, it can become clear that the official raison d'être assigned by the DEM to a certain domain of school mathematics presents limitations, contradictions or incompleteness.

1.2 The phenomenotechnical role of reference epistemological models

A DEM of a certain domain of the mathematical knowledge that is taught (in a given institution) is characterised by not only by the type of mathematical activities that are carried out in that institution around that domain but also by the possible ways of carrying out its study. Consequently, research in didactics must explain why a particular DEM exists, how it determines the structure and functions of the instructional formats and how the emerged didactic phenomena depend on the nature of the specific DEM. At the same time, the criteria used by didacticians to construct an alternative REM to the observed DEM come from a detailed analysis of a set of didactic facts supposedly conditioned by the DEM, which in turn, will be interpreted with the help of the main features of the REM that is being constructed.

Paradigmatic examples of REMs are those proposed by the theory of didactic situations in terms of *fundamental situations* of a given piece of mathematical knowledge (Brousseau, 2002). In the case of the ATD, REMs usually appear as a tree structure of praxeologies. In general, we can say that most of the REMs play a phenomenotechnical role in Bachelard's sense. Their construction gives visibility to certain didactic phenomena and provides tools to formulate didactic problems related to these phenomena. At the same time, they are at the basis of the design of teaching and learning sequences to address these problems and test the hypotheses underlying the REMs (and the related DEMs).

Thus, for example, Francisco Javier García, Josep Gascón, Luisa Ruiz-Higueras and Marianna Bosch (2006) describe the didactic phenomenon of the school isolation of proportionality concerning the universe of elementary functional relations. To this end, it was necessary to situate the teaching of proportionality within the scope of a much broader problem—the connection of school mathematics—and to reformulate mathematical modelling as a connecting instrument. The proposed REM is about not only proportionality but also the modelling of systems of variation between quantities. In this context, proportionality appears as a particular model (among others) of functional relations between magnitudes. This REM makes it possible to account for the isolation of proportionality in school mathematics and formulate and study some didactic problems related to the reduction of proportionality to a purely arithmetic relationship. One of them is the archaic survival of components of the classical organisation (such as the "rule of three"); another is the avoidance of algebraic techniques in the treatment of school problems of proportionality. As for teaching practices, the proposed REM supports a didactic organisation aimed at overcoming the limitations and contradictions of the current school study processes around proportionality. Similarly, diverse investigations of our group have brought to light and allowed the study of other didactic phenomena in different areas of school mathematics. Among these, we can mention the following: Cid, Bosch, Gascón and Ruiz-Munzón (2017) in the field of negative numbers, Barbé, Bosch, Espinoza and Gascón (2005) for function limits, Barquero, Bosch and Gascón (2013) in relation to mathematical modelling, and Ruiz-Munzón, Bosch and Gascón (2011) with regard to elementary algebra.

We can summarise the main characteristics of a REM related to its phenomenotechnical function in the following points:

1. A REM is constructed in didactics of mathematics as a heuristic tool, with methodological objectives, to make certain didactic phenomena visible. Its first function is to provide the necessary elements to formulate didactic problems whose study will improve the knowledge of these phenomena. Only in this way can didacticians emancipate themselves from the DEMs and autonomously construct their objects of study.
2. A REM is not simply associated with a domain of the mathematical activity, but rather with one or more didactic phenomena involving a more or less extensive scope of mathematics. Therefore, the same domain of the mathematical activity can be constructed in different REMs to study diverse phenomena.
3. In scientific practice, the construction of a REM, the empirical analysis of the DEM in the considered teaching institution and the explanation of the associated phenomena are simultaneous processes dialectically developed.
4. REMs are a way for researchers to make their epistemological assumptions, about the subject matter that is taught and learnt, explicit. This explicitness is essential not only to formulate didactic phenomena as research problems but also to expose, control and be able to change them.
5. All REMs are provisional because they are *scientific hypotheses* to be empirically tested. Experimentations based on a REM always provide elements for its evolution. Moreover, if a specific REM does not adequately fulfil its phenomenotechnical function, it must be reviewed and even profoundly modified.
6. To test empirically whether a REM fulfils its heuristic function, and to decide between two rival REMs, didacticians can use the methodology of didactic engineering through the experimentation of teaching proposals supported by the REM to be evaluated.

The next section illustrates the interdependence between the analysis and interpretation of the DEM of a certain domain of school mathematics; the construction of an alternative REM of that scope; and the emergence of certain didactic phenomena. It will end up referring to the nature of REMs as scientific hypotheses and to the empirical testing of these hypotheses.

2 The case of elementary differential calculus and its role in functional modelling

In what follows we discuss in some detail the heuristic phenomenotechnical function of a REM built around the relationship between *functional modelling* (FM) and *elementary differential calculus* (EDC) in the transition between secondary education and university in the educational systems of Portugal and Spain (Lucas, 2015). For the sake of brevity, we will simply call "reference institution" the school institution that is situated in the transition between secondary and university education in the educational systems of Portugal and Spain.

2.1 A sketch of the dominant epistemological model

We will start with two sets of questions Q_1 and Q_2 posed to the reference institution. Portuguese, French and Spanish textbooks, written to correspond with official curriculum guidelines, constitute our empirical basis to interpret the DEM around the FM and the EDC in the reference institution (for more detail see Lucas, 2015 and Lucas, Gascón & Fonseca, 2017).

> Q_1: What is the official raison d'être of the EDC praxeologies in the reference institution? In other words, what types of tasks require the use of the techniques and the technological-theoretical discourse of EDC?

> Q_2: How is FM interpreted in the institution and what role does EDC play in it? What type of tasks can be considered as parts of a FM process?

We summarise below the elements of response that are essentially common to both Portuguese and Spanish educational systems. We begin with the answers to Q_1.

According to textbooks and curriculum guidelines, the official raison d'être for the derivative of a function at a point is centred on the calculation of the limit of the incremental quotient and the slope of the tangent line. Its main aim is to identify critical points of a function, as given by its algebraic expression $f(x)$, and to sketch its graph. The calculation of primitives makes sense in order to calculate the definite integral of a function. The tasks that justify the study of definite integrals are the calculation of areas of flat regions, volumes and lengths of curves. Therefore, definite integrals are essentially used as algorithms to calculate the measure of a quantity.

The responses to Q_2 can be summarised as follows. The functional models that appear in school mathematics are given by *isolated functions* (not by *families of functions*). For instance, students are required to sketch the graph of functions like $f(x) = 0.2x/(1 + x^2)$ or $f(x) = (x + 1)^2 e^{-x+3}$, but they do not systematically study rational functions $f(x) = P(x)/Q(x)$ depending on the degree and roots of polynomials $P(x)$ and $Q(x)$, or the product of a polynomial by an exponential function $f(x) = P(x)e^{Q(x)}$, etc. Moreover, the tasks and techniques used rarely include the construction of functional models. They mostly consist in transforming a model given as a function, and in interpreting the results obtained in terms of the modelled system. Functional processes are very much guided. They are structured in a linear series of small-chained questions limiting the autonomy of the student. The few functional models that appear play the role of applications of previously studied functions with the aim of using their properties and practising the techniques involved. Here are two examples reproduced from

Portuguese textbooks (Viegas, Gomes & Lima, 2011, Tema II, p. 169; Carvalho e Silva, Pinto & Machado, 2012, vol. 3, p. 106):

> PROBLEM 1. The concentration C of a given drug in a patient bloodstream t hours after injection is given by the expression:
>
> $$Ct = \frac{0,2t}{1 + t^2}$$
>
> Calculate $C(1)$ and $C(12)$.
>
> Calculate $\lim_{t \to +\infty} C(t)$ and interpret the result.
>
> PROBLEM 2. Let us suppose that the quantity, B, of thousands of bacteria in the water of a swimming pool from the moment a treatment starts is given approximatively by:
>
> $$B(t) = (1 + 1.5)^2 e^{-0.55t+3}, (t \geq 0)$$
>
> Variable t indicates the time, in days, from the starting of the water treatment. Determine the number of days needed for the water treatment to become efficient, that is, for the number of bacteria to start to decrease.

More generally, in school mathematics, the relationships between discrete and continuous models are not studied. There is almost no work with finite difference equations. Therefore, one of the possible raisons d'être of derivatives, the technical economy they provide compared to discrete techniques, hardly exists. No functional models are built from raw discrete data neither from discrete variational data (obtained either from the average rate of change or from the relative average rate of change).

In summary, our study of Spanish, French and Portuguese textbooks and curriculum guidelines show that the official raison d'être of the EDC components is mainly constituted by a disjoint set of problems and tasks related to the calculation of measures of quantities and to the formal study of given functions. Some of these types of tasks could be considered as components of potential modelling processes, but modelling processes are not really taught, they are absent from the curriculum.

2.2 Proposal for an alternative reference epistemological model

To overcome the limitations suffered by the study processes supported by the DEM around EDC and FM at the reference institution, we propose an alternative REM that fulfils certain conditions. The limitations (and the criteria for deciding what conditions the REM must comply to overcome them) are obtained by the praxeological analysis of school mathematics. In the present case, Catarina Lucas (2015) and Catarina Lucas, Josep Gascón and Cecilio Fonseca (2017) propose the following components of FM processes that give a possible raison d'être to EDC. These components are represented in Figure 6.2.

The first stage of the FM process includes the formulation of a *generating question* Q_0 *about a system (or a piece of mathematical or extra-mathematical reality) and the delimitation of the system.* For instance, a possible question can be to forecast the number of affected persons of a given outbreak when we know those affected in the n first periods (see Figure 6.1). The system is delimited through the election of some factors represented by quantitative variables and some hypotheses about them. In our example, the variables are the time and the number of affected persons. The data given include some implicit hypotheses about the sufficiency of the information available and the chosen time period (years instead of minutes or days).

The second stage consists of the *construction of the model.* Depending on whether the considered variables are discrete or continuous, different types of models can be considered. In the discrete case, the first model is the numerical or graphical one given by the initial raw data (number of infected people). Depending on the values obtained, one can decide to work directly with this variable, with its average rate of change (ARC) or with the relative average rate of change (RARC).

If we decide to remain in the discrete case, the modelling process continues with the formulation of hypotheses about the rates of change and the use of algebraic formulas to describe them. We obtain a model in finite differences that can be sometimes solved algebraically or might require the use of differential tools (back to the continuous approach). In our example, an option can be to always forecast a given percentage of the previous value, obtaining a very simple model of the type $X_t = rX_{t-1} = r^t X_0$. If we decide to obtain a functional model (continuous case), different types of regressions can be considered and compared, until the selection of the family of functions that seems best according to the initial question, the variables and hypotheses (a straight line, a parabola, an exponential, etc.). The resulting model can be a differential equation (if the initial

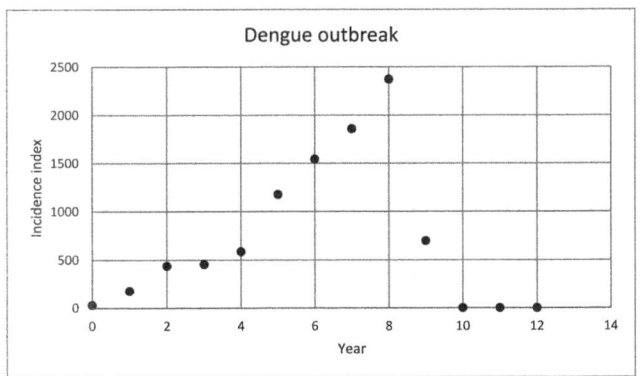

FIGURE 6.1 System given by a set of numerical data and a graph

variables are ARC or RARC) or a functional one. In the example, one can find that the RARC can be fitted with a straight line of equation $y = 2.3 - 0.4t$ where $y = (X_t - X_{t-1})/X_t$. In some cases, instead of working with raw data, one can directly use "exact" models given by a functional expression or "differential" models given by a differential equation.

The third stage of the FM process corresponds to the *work within the model* and the *interpretation of the results obtained* in terms of the initial system, the meaning of the model parameters, the comparison between different possible models and outputs, their limitations (in our example, possible differences between the short or long term), etc. Finally, the fourth stage includes *raising new questions* and the need to *extend the functional modelling process* (considering new variables, new hypotheses, etc.).

In summary, the REM about FM processes will cover the different stages of the mathematical modelling process: delimitation (or construction) of the system; model construction; model work and interpretation of the results in terms of the system; formulation of new problematic issues, new hypotheses and the beginning of a new FM process. It also considers relationships between discrete and continuous functional models. Finally, the REM emphasises the role of EDC as a tool not only to work with functional models and evaluate them but also to construct the models according to the initial situation and related questions. This includes, in particular, the interpretation of the parameters of the model in terms of the modelled system.

Figure 6.1 presents a scheme of the four stages of the REM. It includes different processes of construction, use and comparison of functional models, the distinction between the discrete and continuous case, the relationships between them and the roles of EDC therein. We will explain it by distinguishing different cases as shown in Figures 6.2–6.5.

Figure 6.3 corresponds to the discrete case. The starting point is a generating question about a system which is initially delimited by selecting some discrete data of some of its variables (1) and (2). The system can be initially described by a *numerical model* (raw data table) and plotted (*graphical model*) (3). The variation of the corresponding variables can then be studied by considering a table of values of their average rate of change or of their relative average rate of change (4). Some hypotheses can be formulated about the types of elemental variations considered (5). If any of the hypotheses is confirmed, then it is possible to write a *variational algebraic model* in the form of a finite difference equation (6), the resolution of which (7) leads to a *discrete functional algebraic model* (8). Working within this model can bring answers to the initial questions (9) and interpretations in terms of the initial system (10). This could lead to introduce other variables, formulate new hypotheses (11) and restart the modelling process (1).

If the data are continuous (Figure 6.4), instead of a functional-algebraic model one can construct a *differential model* (3) whose integration (4) leads to a *continuous functional model* (5) and keep working on it in the continuous field (6–9).

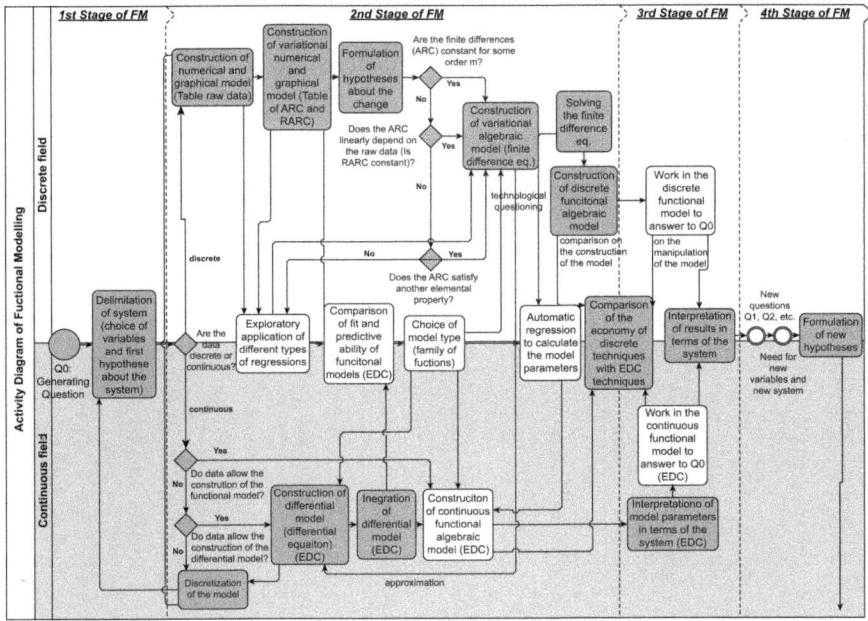

FIGURE 6.2 Structure of the activity diagram that redefines functional modelling

Figure 6.5 identifies a path that starts from discrete data and assumes that no elementary functional model (polynomial, exponential, logistic, etc.) fits it. In this case one can find technical difficulties to work with finite difference equations (6). Other types of regressions can be applied (5) on the variational data (ARC or RARC) (4). These regressions are approximations of the *discrete variational models* (6), which, in turn, will be approximated by *continuous differential models* (7). Similarly, as before, one can solve the associated differential equations to obtain families of functional models as solutions (8). When comparing the fit and predictive capacity of the different models, one can choose the best model of each family (9) and the most appropriate to describe the system (10). The last part of the path (11–15) is similar to the previous one (Figure 6.4, 5–9).

In summary, we want to underline that, as we have said before, the alternative REM we propose around FM assigns a nuclear role to the EDC. This role is not very visible when the FM process is located entirely in the discrete field (Figure 6.3), but if the data available are continuous (Figure 6.4) the techniques of the EDC are fundamental to construct the functional model, work on it and interpret the results of this work in terms of the system. In scientific practice, it is very common that, even starting from discrete data, it is necessary to work in the continuous field. Indeed, when starting from discrete data (Figure 6.5) it may be profitable, for reasons of technical economy, to approximate the models in finite differences using a continuous differential model. This underlines the importance of the role of EDC techniques in the vast majority of functional modelling processes.

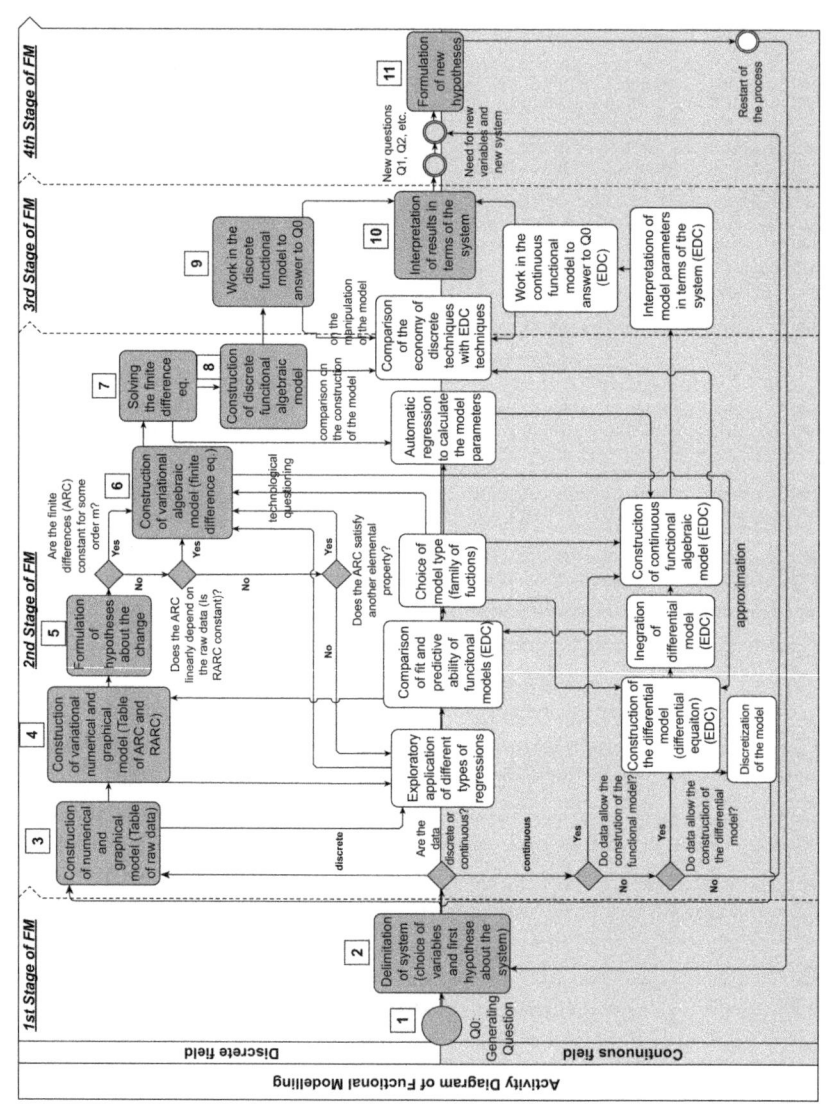

FIGURE 6.3 Example of a functional modelling process in the discrete field

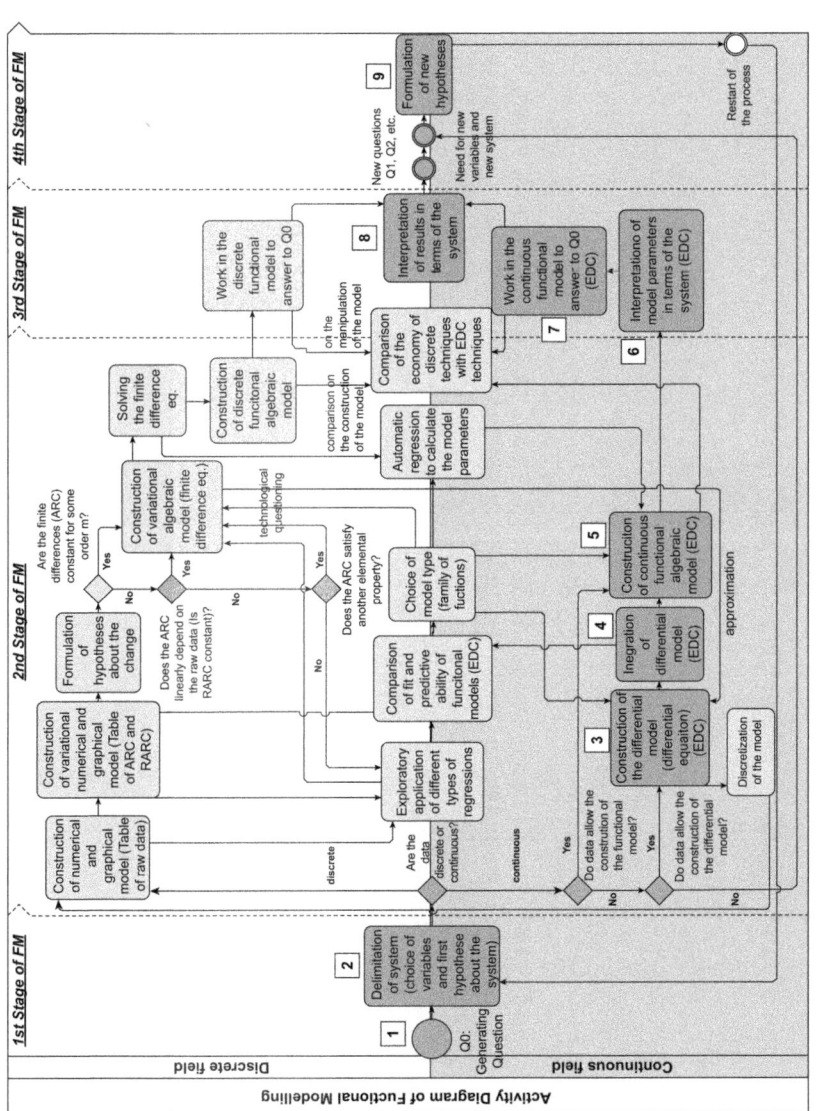

FIGURE 6.4 Example of a functional modelling process in the continuous field

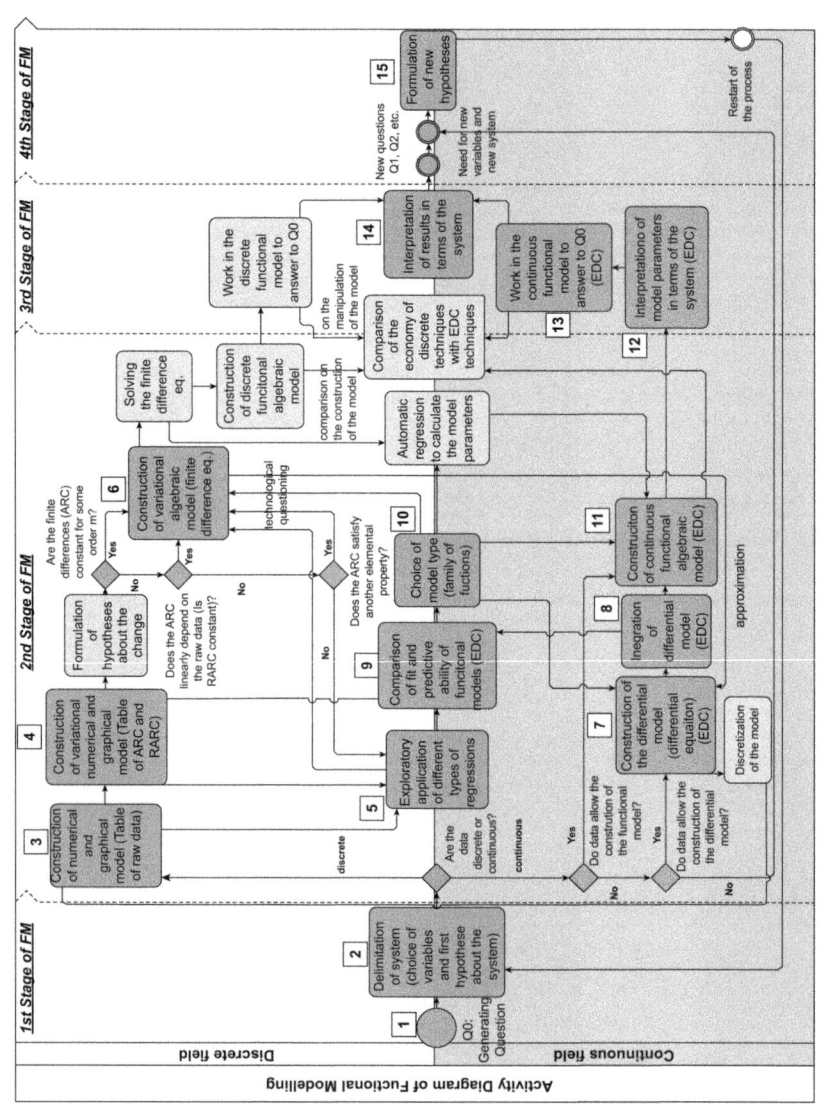

FIGURE 6.5 Example of an MP that allows the step from discrete to continuous

2.3 The use of the REM to analyse school mathematics

The proposed REM assigns an alternative raison d'être to EDC, compared to the official one, consisting of problematic questions and tasks that lead to the development of FM processes. In these processes, the use of EDC techniques is more motivated by their operational economy when working with functional models (compared to the algebraic techniques) than by conceptual needs related to the notions of limit and infinitesimal variation.

The modelling processes that compound the REM are practically absent in the school institution of reference. We can show and visualise this fact in the activity diagram by shading the components of the modelling processes that, in one way or another, are present in the school mathematical practices (see Figure 6.6). The absence is especially evident if we consider the processes of functional modelling that are based on this diagram as a whole, that is, if we treat each path as an indivisible one that starts from a problematic question and culminates in an answer to that question. We can see that the tasks proposed by the educational system as the official raison d'être of the EDC are quite isolated from each other and not explicitly related to the FM.

By highlighting the limitations of the official raison d'être of EDC, the alternative REM gives visibility to some aspects of various didactic phenomena, which are closely interrelated. First, it appears as a particular case of the general phenomenon of rigidity and atomisation of school mathematical praxeologies described by Cecilio Fonseca, Marianna Bosch and Josep Gascón (2004). In Lucas (2015) we study this phenomenon by investigating the historical evolution of the role of EDC and its relationship with FM in the reference institution, the origin of this phenomenon in didactic transposition processes, the conditions that maintain it and its main didactic consequences.

Closely related to this phenomenon is what we can call the "poverty" (in terms of a variety of mathematical tasks) of the school activities around FM and the corresponding lack of visibility of modelling activities. The logical consequence is that EDC's raison d´être cannot be related to the construction, manipulation and interpretation of different types of functional model. The tasks related to solving optimisation problems, which are dominant in school mathematics, are a special case of the study of a system from a functional model. However, in these tasks, the role of EDC is essentially restricted to the calculation of the extremes of the functional model.

We can even talk of hidden functional models in school mathematical activities, since some types of tasks proposed to give sense to the study of EDC (for example, the calculation of primitives or definite integrals, the sketch of functions given by their analytic expression, etc.) are not recognised as components of FM processes. To sum up, we can affirm that the raison d'être the REM assigns to EDC in the scope of the FM is a radical extension (or generalisation) of the one assigned to it by the current educational system, more coherent with the role EDC plays in many scientific activities. Therefore, the REM relates the didactic

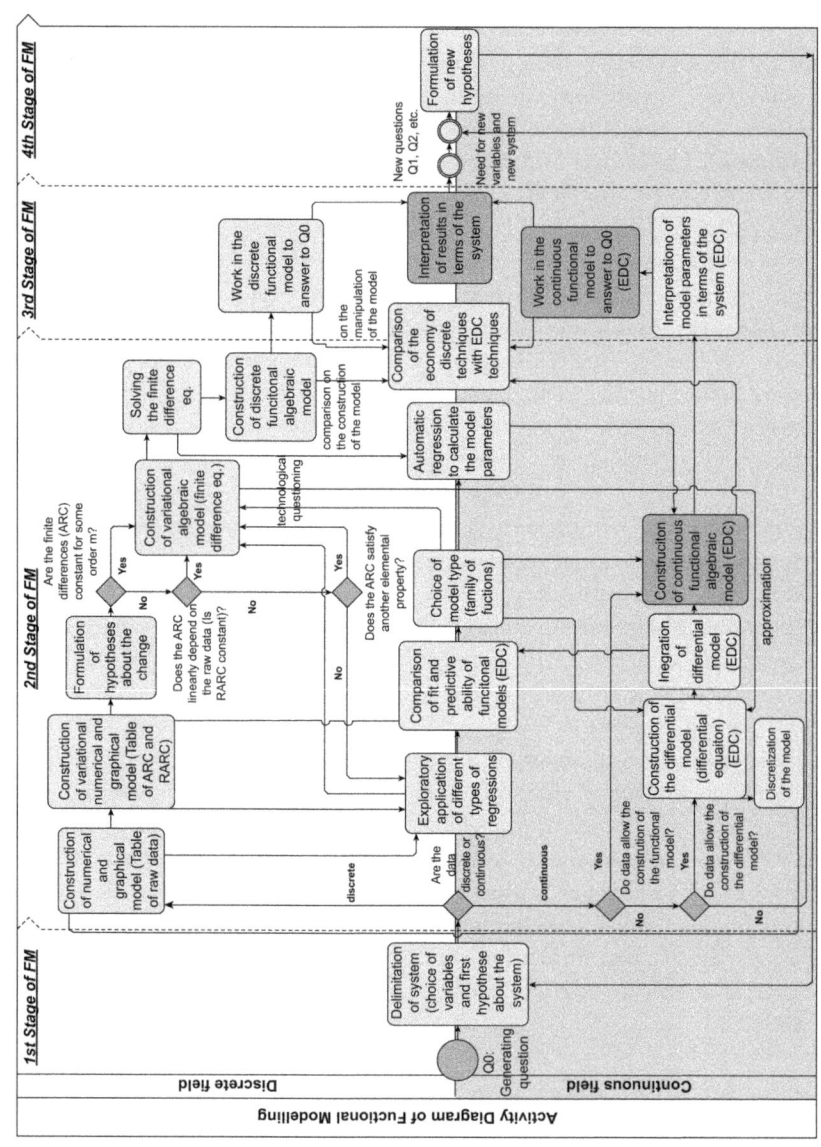

FIGURE 6.6 Components of the activity diagram of functional modelling worked in the reference institution

phenomena related to the teaching of EDC at secondary education to the irrelevant role FM still has in secondary education and the transition to the university.

This specific REM can be interpreted as a scientific hypothesis that can be briefly formulated as follows. If under certain conditions, a study community carries out a study process sustained in that REM, then that study community will carry out a mathematical activity that will not suffer from the limitations of school study processes supported by the DEM. This hypothesis has been partially contrasted through the experimentation carried out with Portuguese university students of the first year of Nuclear Medicine (Lucas, 2015). In effect, these students carried out various FM activities in which the four stages of the modelling process appeared to answer the generating question. Students worked with discrete or continuous models (depending on the nature of the data and the modelled system) and, in all cases, EDC not only provided calculation techniques but was also used to systematically build the differential models obtained.

3 A second REM to connect the construction of continuous models of physical and geometrical quantities to the study of parametric functional models

This section presents a different type of REM that was initially built to take into account proven learning obstacles in elementary differential calculus. It is elaborated from a more generic REM that includes two big types of praxeologies, "modelling" and "deductive". It originated in the research laboratory in didactics of mathematics, the Ladimath, led by Maggy Schneider at the University of Liège.

3.1 Learning obstacles related to problematic preconstructs

The thesis and projects that fall under the umbrella of the Ladimath concern a category of phenomena related to the relevance of certain mathematical concepts for modelling preconstructions in the sense of Chevallard (1985/1991) in his theory of didactic transposition (Chevallard & Bosch, 2014) or for modelling mental objects in the sense of Hans Freudenthal (1973). For example, high school students and even mathematics teachers in training have a general position towards mathematics that comes close to *empirical positivism*. In the domain of sciences epistemology, empirical positivism means a perception of scientific laws and concepts as an exact reflection of objects in the "physical" world, not as human intellectual elaborations aiming at specific objectives. The result of such a position would be a lack of detachment between the observed "phenomenon" and the concepts enabling to develop an appropriate model (Job & Schneider, 2014).

Such a vision of mathematics allows explaining errors related to the *obstacle of dimensions heterogeneity* (Schneider, 1988, 1991a) as unconscious shifts from quantities to their measures—for example, when someone assumes that the volumes of two revolution solids are to each other as the areas of their generating surfaces. We may then observe that students justify this false result by using arguments involving quantities, insisting on the fact that solids are composed by their radial sections. Beyond this narrow context, this positivist conception also allows the interpretation of reactions referring to geometrical objects or quantities defined by the concept of limit. As an example, the tangent may be defined as a "limit" of secants around a point, without any former knowledge of a topology on the set of lines, rather than as a line defined using its slope or as a limit of functions in the usual mathematical sense. This is the so-called *geometrical obstacle of the limit* (Sierpinska, 1985; Schneider, 1988, 1991b). This obstacle also appears in the case of the area of curvilinear figures (Schneider, 1991a) and its calculation through a sequence of areas of rectangles, as "taking the limit" is often conceived in terms of rectangles being reduced into segments, and therefore of area 0. Students act as if the limit was directly carried out on geometric objects and not on the measures of these objects. Regarding now the instantaneous velocity, pupils usually claim that it cannot exist as it cannot be exactly assessed using observations and physical measures. In that sense, they actually refer implicitly to a sensible universe and not to the world of concepts elaborated by human minds (Schneider, 1992).

As Pierre Job (2011) has shown, this empirical obstacle persists well beyond this context. It also pervades the relations pupils have to definitions at the end of secondary education. For them, definitions are "non-lakatosian": they are supposed to describe the intuitive meaning of "things" to be defined, rather than being *models* of these "things", designed to serve as foundation for deductive architectures. From an institutional point of view, we can postulate that these preconstructions are not born ex nihilo and that epistemological obstacles have deep roots in culture. In particular, in a positivistic vision in which "modern" science is presented as an "act of reading nature" in order to understand its intrinsic laws.

Sometimes the preconstructs are the consequence of previous education. A typical example is the tangent defined in a certain way in the case of the circle, which will be in blatant contradiction with what will be later called in analysis the tangent at a point of a curve (Schneider, 1991b; Castela, 1995). Therefore, the origin of learning obstacles is sometimes epistemological, sometimes didactic. In both cases, they require a background treatment such as the one we describe below.

3.2 Two praxeological levels and fundamental situations in a broad sense

From these institutional analyses, especially the ones about the empirical obstacle, we conclude that there is a need for two praxeological levels: one taking into account preconstructed knowledge and another its scientific construction. From there, Maggy Schneider (2008) distinguishes two types of mathematical praxeologies, a "modelling" one and a "deductive" one in the sense of processes to describe two aspects of mathematical activity and also in the sense of the products of these processes. In the "modelling" type, the goal is to model objects that are not mathematically defined but of which one has a certain knowledge (they are "preconstructs"). It implies relying on geometrical or kinematic arguments like these: the measures of nested surfaces respect a certain relationship of order (the ones "inside" being lower than the ones "outside"), and a particle that accelerates from one speed to another passes through all the intermediate speeds. It also implies the use of pragmatic validations—that is, testing the reliability of a new technique on problems already solved with another trusted method. For example, ensuring that a limit calculation provides the "exact" value of a curvilinear area or the "exact" value of an instantaneous velocity. In the "deductive" type one builds up a deductive organisation from the properties of the thereby constructed model, the objects being defined by the techniques that were used to model them in the previous "modelling" praxeology. The validation follows the rules of deductive logic giving birth to definitions able to support deductive reasoning. For example, the instantaneous velocity is defined as a limit.

In passing from one praxeological level to another, one must consider the importance of fundamental situations, in the broad sense (Schneider, 2008), whose major stake is not that much in the construction of a specific knowledge by the students, but more in them entering a new institution where the relation to an object of knowledge is different. For example, the evolution of geometry from lower to higher secondary school, which does not consist of describing what one sees but of establishing what must be seen. Or, the need to change students' conception of definitions, from infinitesimal calculus to mathematical analysis, as explained above (Job & Schneider, 2014).

The division of the mathematical activity into two articulated praxeological levels, together with the extension of the concept of fundamental situation, characterises the adopted REM. Let us now see how it materialises in the case of mathematical analysis and its phenomenological role.

3.3 The two-component REM and its phenomenotechnical role

The first praxeology, of the modelling type, leads to the infinitesimal calculus. It concerns indeed the determination of quantities or geometrical objects within mathematics or other domains like geometry or kinematic: curvilinear areas and volumes, variable velocities, tangents. These are the fundamental tasks of infinitesimal calculus together with questions of the optimisation of quantities. In this praxeology, quantities are not really defined and are used as a kind of preconstruct. The calculation of limits will here appear in an "embryonic" form as a technique consisting of suppressing specific terms in an expression obtained after the standard simplifications and without compensation, as is usual in algebra. For instance, the average rate of change of the law of motion $p(t) = t^2$ is $[(t + \Delta t)^2 - t^2]/\Delta t = 2t + \Delta t$ and we delete Δt to obtain $v(t) = 2t$.

Other techniques, like the usual calculation of derivatives or primitives, quite rapidly supplant this one. In this first praxeology, the so-called technological discourse has a very specific objective. It does not consist of demonstrating results on limits calculation in an algebraic structure but aims at validating that such a calculation technique provides the "exact" value of a curvilinear area or an instantaneous velocity. As seen above, this implies using geometrical or kinematic arguments and/or pragmatic validation.

This first praxeology will be called "quantities modelling" and should be connected to another praxeology, also of the modelling type, called "functional modelling". Its major task consists in classifying various phenomena (in mathematics or not) using parametric functional models. A second praxeological level, of the deductive type, corresponds to the elaboration of the mathematical analysis, independent from geometry or kinematics. It is based on a fundamental situation in the broad sense described in Job (2011) and will not be developed here.

A first phenomenotechnological role of this global REM consisting of two praxeological levels (modelling and deductive) has been to describe the learning obstacles mentioned above. A second phenomenotechnical role appears in the use of the REM to characterise the dominant epistemological model (DEM) in secondary school institutions, which does not allow taking into account these same learning difficulties and neglects the embryonic forms of knowledge, more or less distant from socially standardised forms. In Belgium, this DEM concerns secondary education and didactic devices for the transition to higher education. The teaching in secondary schools focuses on the use of techniques accompanied by isolated items of the academic teaching's formalism. Modelling appears neglected, or at least treated in a very allusive manner. The functions are studied for themselves and, most often, one by one in the absence of parametric models. Finally, we could say that this teaching sits alongside the study of procedures and some "serious"-looking theoretical results, believing that it is a good preparation for further studies—for example, at university (Rouy, 2007).

4 As a synthesis: institutional emancipation and phenomenotechnical character of didactics

Every science has the ambition to constitute its *object of study* in a relatively autonomous way. This construction of the object of study can be identified with the *production of phenomena* and of *research problems* associated with the study of these phenomena that constitute its raison d'être. In fact, we consider that all sciences (and in particular social sciences), according to their degree of development, have the potential capacity to produce the phenomena they describe and explain. In other words, we postulate that all sciences potentially have a *phenomenotechnical* character in the sense of Bachelard and that this character is related to the degree of *emancipation* achieved in each historical period.

The *institutional emancipation* of didactic science can be interpreted as the liberation from the conditions imposed by educational institutions—that is, of the prevailing codes and, in particular, of the way they formulate and interpret the questions that are part of the scientific problem. In particular, the REMs, at the core of the ATD methodology, play a key role as instruments of *epistemological emancipation* of didactic science and, simultaneously, have an important phenomenotechnical function. They strongly condition the didactic phenomena that will be "visible", the description and study of those phenomena, the attempts to explain them, the engineering projects that may cause the emergence of certain phenomena, and the teaching and learning processes that can be proposed to modify the usual school didactic organisation.

In the case of the elementary differential calculus (EDC), we have described two different proposals of REM, together with their phenomenotechnical functions. The first REM locates the EDC within the scope of functional modelling (FM) and assigns it an alternative raison d´être that allows us to reveal certain didactic phenomena, such as the school disconnection between FM and EDC, the lack of visibility of FM activities and the existence of hidden functional models. We can consider that these phenomena are associated with obstacles (difficulties or constraints) that arise at the *disciplinary level* in the *scale of levels of didactic codeterminacy* (see Glossary), or at most at the *pedagogical level*. They are related to the rigidity and atomisation of all school praxeologies and, in particular, of mathematical praxeologies.

If we expand the universe of our analysis and look at the epistemological and didactic obstacles that arise in the transition from EDC to *mathematical analysis*, we need a more comprehensive REM to approach didactic phenomena that emerge at the upper levels of the scale, including the *level of societies* and *civilizations*. The prevalence of *empirical positivism* as spontaneous epistemology can explain general phenomena that manifest as obstacles related to the confusion between the *studied system* (which can be physical, geometrical, economical or of any other type) and the *mathematical model* that is constructed to study it. The consequence is obstacles such as the geometrical obstacle of the limit, the heterogeneity of dimensions, and the descriptive character assigned to definitions. They appear when mathematical models can exist by themselves and crystallise in new objects that become independent of the systems initially modelled. The second type of REM that we propose to explain the learning difficulties associated with these obstacles consists of two articulated praxeological levels: a "modelling" type (which, in turn, articulates the "modelling quantities" with the "functional modelling") and the "deductive" type. This two-level REM proposes a way of interpreting scientific knowledge that overcomes the *empiricist prejudice* of knowledge coming directly from a supposedly universal empirical reality. By presenting scientific knowledge as mediated by the construction of models and of "models of models", this REM highlights the didactic phenomena associated with the transition between elementary differential calculus and mathematical analysis. Pursuing the phenomenotechnical work initiated with these two REMs requires new research to design and implement new instructional proposals based on them as a way to test their potential, to refine and develop them, and to propose new adaptations according to different school levels, pedagogical environments and institutional settings.

References

Bachelard, G. (1949). *Le rationalisme appliqué*. Paris: Presses Universitaires de France.

Barbé, J., Bosch, M., Espinoza, L., & Gascón, J. (2005). Didactic restrictions on the teacher's practice: The case of limits of functions in Spanish high schools. *Educational Studies in Mathematics*, 59(1–3),235–268.

Barquero, B., Bosch, M., & Gascón, J. (2013). The ecological dimension in the teaching of mathematical modelling at university. *Recherches en Didactique des Mathématiques*, 33(3), 307–338.

Brousseau, G. (2002). *Theory of didactic situations in mathematics*. Dordrecht, The Netherlands: Kluwer Academic Press.

Castela, C. (1995). Apprendre avec et contre ses connaissances antérieures. *Recherches en Didactique des Mathématiques*, 15(1), 7–48.

Carvalho e Silva, J., Pinto, J., & Machado, V. (2012). *NiuAleph 12: Manual de Matemática para o 12.º ano de Matemática A*. Vol. 3 (edição de autor).

Cid, E., Bosch, M., Gascón, J., & Ruiz-Munzón, N. (2017). Integración de los números negativos en el modelo epistemológico de referencia de la modelización algebraico-funcional. In G. Cirade *et al.* (eds), *Évolutions contemporaines du rapport aux mathématiques et aux autres savoirs à l'école et dans la société* (pp. 325–341). Toulouse, France: Université Toulouse Jean Jaures.

Chevallard, Y. (1985/1991). *La transposition didactique: Du savoir savant au savoir enseigné*. Grenoble, France: La Pensée Sauvage (2nd edition 1991).

Chevallard, Y., & Bosch, M. (2014). Didactic transposition in mathematics education. In S. Lerman (ed.), *Encyclopedia of mathematics education* (pp. 170–174). Dordrecht, The Netherlands: Springer.

Fonseca, C., Bosch, M., & Gascón, J. (2004). Incompletitud de las organizaciones matemáticas locales en las instituciones escolares. *Recherches en Didactique des Mathématiques*, 24 (2/3), 205–250.

Freudenthal, H. (1973). *Mathematics as an educational task*. Dordrecht, The Netherlands: D. Reidel Publishing Company.

García, F. J., Gascón, J., Ruiz-Higueras, L., & Bosch, M. (2006). Mathematical modelling as a tool for the connection of school mathematics. *ZDM Mathematics Education*, 38(3), 226–246.

Gascón, J. (1994–1995). Un nouveau modèle de l'algèbre élémentaire comme alternative à l'"arithmétique generalise". *Petit x*, 37, 43–63.

Gascón, J. (2003). From the cognitive to the epistemological programme in the didactics of mathematics: Two incommensurable scientific research programmes? *For the Learning of Mathematics*, 23(2), 44–55.

Job, P. (2011). "Étude du rapport à la notion de définition comme obstacle à l'acquisition du caractère lakatosien de la notion de limite par la méthodologie des situations fondamentales/adidactiques. (Doctoral dissertation). Université de Liège, Belgium.

Job, P., & Schneider, M. (2014). Empirical positivism, an epistemological obstacle in the learning of calculus. *ZDM Mathematics Education*, 46(4), 635–646.

Lucas, C. (2015). Una posible razón de ser del cálculo diferencial elemental en el ámbito de la modelización functional (Doctoral dissertation). Universidad de Vigo, Spain.

Lucas, C., Gascón, J., & Fonseca, C. (2017). Razón de ser del cálculo diferencial elemental en la transición entre la enseñanza secundaria y la universitaria. *REDIMAT-Journal of Research in Mathematics Education*, 6(3), 283–306.

Rouy, E. (2007). Formation initiale des professeurs et changements de posture vis-à-vis de la rationalité mathématique (Doctoral dissertation). Université de Liège, Belgium.

Ruiz-Munzón, N., Bosch, M., & Gascón, J. (2011). Un modelo epistemológico de referencia del algebra como instrumento de modelización. In M. Bosch *et al.* (eds), *Un panorama de la TAD* (pp. 743–765). Barcelona, Spain: Centre de Recerca Matemàtica.

Schneider, M. (1988). Des objets mentaux aires et volumes au calcul des primitives (Doctoral dissertation). Université de Leuven, Belgium.

Schneider, M. (1991a). Un obstacle épistémologique soulevé par des "découpages infinis" des surfaces et des solides. *Recherches en Didactique des Mathématiques*, 11(2/3), 241–294.

Schneider, M. (1991b). Quelques difficultés d'apprentissage du concept de tangente. *Repères IREM*, 5, 65–82.

Schneider, M. (1992). A propos de l'apprentissage du taux de variation instantané. *Educational Studies in Mathematics*, 23(4), 317–350.

Schneider, M. (2008). *Traité de didactique des mathématiques*. Liège, Belgium: Presses Universitaires de Liège.

Sierpinska, A. (1985). Obstacles épistémologiques relatifs à la notion de limite, *Recherches en Didactique des Mathématiques*, 6(1), 5–68.

Viegas, C., Gomes, F., & Lima, Y. (2011). *Xeqmat 11: Manual de Matemática para o 11.º ano de Matemática A*. Portugal: Texto Editores.

PART 3

Questioning the World

7

THE SECOND SPANISH REPUBLIC AND THE PROJECT METHOD

A view from the ATD

Encarna Sánchez-Jiménez, Dolores Carrillo Gallego, Yves Chevallard and Marianna Bosch

1 The Second Spanish Republic

The Second Spanish Republic was the democratic government in office from 1931 to 1939. (The First Spanish Republic had been even shorter-lived: established in February 1873, it ended in January 1874.) The Second Republic was proclaimed on 14 April 1931, after the departure in exile of King Alfonso XIII. Following the bloody three-year Spanish Civil War (1936–1939), which would lead to the long military dictatorship of Francisco Franco (1939–1975), the Second Republic came to an end on 1 April 1939. During the period 1931–1936, and in the previous decades, there was an intense pedagogical renewal in Spain. At that time, innovative currents in teaching were influenced by the New School movement, which emerged at the end of the 19th century and consolidated and spread in the first third of the following century. The New School was an alternative to the traditional one and considered the child as a central element when organising the processes of teaching and learning. Its main characteristic is the importance given to both the activities themselves and the respect for the interests of the children, along with their individuality. It called for a renewal of the education systems and methods and gave rise to many educational experiences worldwide.

Spanish professionals would learn about these methods and experiences through published works and also their travels abroad when they were awarded scholarships by the Extension of Studies Board (Junta para la Ampliación de Estudios). This institution had been created in 1907 to promote the exchange of ideas and experiences among researchers and teachers from Spain and other countries as well as research in Spain and the development of educational institutions. Education was to be one of the pillars of the politics of the Republic, and the improvement of primary school teachers' preparation was a key priority. People related to the New School movement (through the Institución Libre de Enseñanza, the Revista de

Pedagogía, the Revista de Escuelas Normales and the Junta para la Ampliación de Estudios) came to positions of responsibility in the educational sphere during the Second Republic. Through legislation and management they tried to institutionalise the pedagogical ideology and related experiences of the previous three decades in Spain. The new pedagogical methods brought new working methods, individual and collective. Among the latter is the "project method" (PM), a device typifying the international New Education movement (Knoll, 2014).

2 The project method

This method had its influence on teacher education in the *escuelas normales* (teacher colleges), especially during the period of peak renovation in these centres. The *patronatos escolares* (boards of trustees) were founded between 1931 and 1934, during the Second Spanish Republic. They were associated in particular with the *escuelas anejas* (annexed schools), the schools used as practice grounds for teacher colleges, to try out new pedagogical methods. In this way, the number of "research and reform schools" (*escuelas de ensayo y reforma*) increased. The Spanish government had created them in 1922 as institutions to put into practice the ideas of the New Education (Del Pozo, 2003–2004). As a result, trial runs of the project method were introduced into the basic training for teachers.

Three main proponents disseminated the project method in Spain. One was Lorenzo Luzuriaga Medina (1889–1959), a school inspector who founded (1922) and directed the journal *Revista de Pedagogía* (1922–1936), the Spanish organ of the International League of New Education (Viñao, 1994/95). The second one was Fernando Sáinz Ruiz (1891–1957), a school inspector who held a number of positions and became Inspector General of Primary Education. Like Luzuriaga, he took an active part in the educational policy of the Second Spanish Republic. The third proponent was Margarita Comas i Camps (1892–1972), a teacher educator, author of several works on the teaching of mathematics and frequent contributor to the *Revista de Pedagogía*. According to Del Pozo (2009), the three stood halfway between foreign pedagogues who propounded the method and the teachers who experimented with it in their classrooms. Our study is based on the publications of these authors, together with that of David Bayón Carretero, a school inspector who, with the primary school teacher Ángel Ledesma Martín, wrote a book about the experience of putting this method into practice in a school.

Most authors, instead of defining the project method, try to characterise it. Nevertheless, the most commonly accepted definition is the one proposed by John Alford Stevenson (1886–1949), who recognised that the idea had been proposed by Werrett Wallace Charters (1875–1952), professor of education at the University of Illinois, of whom he was a disciple (Stimson, 1954). Margarita Comas translates it as "a simple or complex act, carried out in its *natural environment* as a response to a problem" (Comas, 1931/1963, p. 9). Stevenson (1921) designed a double system of classification for projects: *simple projects* and *complex projects*, on the one hand, and *intellectual* or *manual projects* on the other hand. Regarding the division of projects between intellectual and manual projects, Fernando Sáinz (1928/1961) failed to see a clear boundary between the two. He suggested differentiating projects by subject matter or area. Regarding the classification of simple and complex projects, he provided several examples of the former. Some projects concerned mathematics and were quite close to what would now be considered *problems*: "in language lessons, write letters about an event; [...] in mathematics, find the average pupil attendance in a school, draft a budget for a school outing, find out the number of tiles necessary to fix the floor" (Sáinz, 1928/1961, p. 35, our translation).

However, those who defend the project method place emphasis on distinguishing between a *project* and a *problem*. That distinction depends not so much on size, scope or content, but rather on function. Problems are considered a part of projects if they are not simple enough to require the application of known principles exclusively. In the paper "El Método de Proyectos en la Enseñanza" ("Project Methods in Teaching"), published by the journal *La Escuela Moderna* between 1933 and 1934, a theory in this respect was proposed by William Heard Kilpatrick (1871–1965):

> It is obvious that each accepted problem, whose solution requires a natural process, is ultimately a project. Yet not every project is a problem. The problem method becomes, thus, a special case, surely one of the most important ones in the project
>
> *(La Escuela Moderna, 1934, p. 129, our translation)*

3 The reception of the project method

The project method is interpreted as a solution to organisational matters and curriculum issues in schools that had been organised on the principles of the Escuela Nueva (New School) movement. It is informative to determine the questions that were raised and the didactic organisations that were built in response (Sánchez-Jiménez, 2015). F. Sáinz (1928/1961, p. 32) posed the following question, which could be considered the *driving question* behind the method:

> Q_0: Is it possible [how] to organise a school following a task plan that is equivalent to the one developed outside the school, i.e. at home, in the streets, in society?

The project method is an answer to that question, and leads to a more specific one:

> Q_1: How can the project method be used? What part of the school activity should be developed with this method?

A number of ways to put this method into practice were found. It could be applied to every school activity, and the whole school curriculum could be taught using projects. There were also intermediate uses. Considering the existing institutional conditions, Fernando Sáinz proposed a moderate use. Thus, he suggested using this method: (a) only for some subjects in the school curriculum; (b) at specific points in the timetable each day or on specific days every week; (c) with a certain year; or (d) with a selection of students, e.g. with the most advanced students.

The other question that comes up is:

Q$_2$: How can the project(s) be selected? Which projects should be chosen?

This question can be broken down into the following ones:

Q$_{21}$: What "breadth"—when compared with the school curriculum for one year—should a project have?
Q$_{22}$: Which criteria should be used to select a project?

The first discrepancy that arose between the founders of this method concerned who should come up with the project. Some of them held that only the teacher was in a position to have some guarantee of success. Others contended that the project must come from the children themselves, spontaneously. Sáinz (1928/1961, pp. 36–37) adopted an intermediate position, admitting that it was important to take the children's proposals into account but advising teachers not to accept a project that did not comply with certain specifications, even if the class had suggested it.

The response by Margarita Comas to the question came as a list of queries that the teacher must consider when designing or filtering projects:

1. Is the project interesting for most of the group?
2. Does it have enough value so that individuals can provide a definite contribution to their own development or the development of the group?
3. Does it open individuals, either consciously or unconsciously, to new horizons where they can find new problems to solve, and thus new projects to undertake?
4. Does it help to enlighten any phase of the experience or activity of the child, that might be worth defining and preserving, even temporarily?
5. Does it gradually expand the children's interest in something and bolster their capacity for sustained attention?
6. Could this project, better than others, solve some problem at some point, even through a result that we might consider negative?

(Comas, 1931/1963, pp. 10–11, our translation)

Once a topic related to a focus of interest has been selected, the following *question* arises:

Q$_3$: How should teachers plan and implement the project for their students?

This question takes us to several *didactic types of tasks* for the teacher associated—although not exclusively—with this method, the first being:

T_1: Creating a preliminary plan for the project.

For this task, Stevenson (1921) provided a *didactic technique* that consists in proposing a diagram. In the wake of, for example, A. Rodríguez Mata (1923), this was later expanded by Comas, with an example for experimental sciences:

> Focus of interest [...] Topic [...] School subjects it comprises [...]
> I.   Student's previous experiences that could be tapped
> II.   Teacher's main goals
> III. Different phases in teaching
>   1. Preparing students so they feel the need to learn
>   2. Empowering students so that they can acquire the knowledge they need
>   a) Checking results
>   b) Applying results
> *(Comas, 1931/1963, pp. 11–13, our translation)*

Which aspects should the planning phase include? This outline encompasses the contents of the project and the sequence of activities involved. In addition to this, most authors writing about this method mention the "natural positioning" of the project. This means that the teacher's job also includes making the students feel the need to learn, to be interested in the situation that has been suggested, to accept the challenge, and to make it their own. Students need to understand the result that is required from them and then face the situation as though it were "a match that must be won" (La Escuela Moderna, 1933, p. 566). All this, of course, reminds us of the concept of devolution (Brousseau, 2002).

The way in which projects connect different areas of knowledge and several ways of working is considered one of the factors contributing to children's engagement. However, the teacher must take on other types of tasks, like:

T_2: Handling the execution of the task by the students.

In this case, one of the *didactic techniques* defining project teaching comprises setting up a study-support mechanism, or *didactic device*, which is described in every example. This is *teamwork*. If one of the intrinsic ideas in the project method is children's independent work, under the guidance of the teacher, the other idea is to work in teams. Sáinz considered collective work to be one of the "problems" that the new method solves most efficiently. Children, instead of forming a mass facing a single actor, the teacher, get help from the teacher *and* other students, working as a team.

Defenders of the project method consider that teamwork has several advantages. They normally comment on how the teacher should handle this part of the autonomous work of the teams. The goal of some of the proposed *didactic gestures* is to avoid discipline conflicts during work in teams, or among the teams—for example, by choosing (or asking them to choose) a team representative or promoting some healthy competition among the teams, to foster a spirit of collaboration. Defenders of the project method also provide advice on how to create teams or how to distribute work among them.

They were aware of the institutional situation in which this system would be applied in Spain, and thus they pondered *ecological* questions, taking into account, for example, the fact that schools were not (at that time) divided into grades. They indicated that the teacher can make one group work on a specific project while paying more attention to the rest of the students—either younger or less advanced students—and maintaining the concept of the school as a study community. In one-room schools, teachers must make teamwork part of the school life, under the consideration that:

> Schools cannot lose their superior unity, and the group or groups must always try to communicate with their classmates, tell them about their work and transmit the feeling of being exemplary in their work. Also, children's individual responsibility must not be lost in the group or in anonymity.
>
> *(Sáinz, 1928/1961, pp. 65–66, our translation)*

Another didactic type of tasks teachers must carry out is related to the validation and institutionalisation of knowledge:

> T_3: How to promote [organise, generate, prompt] the validation and institutionalisation of the knowledge managed?

As mentioned in the scheme proposed by Comas, validation actions normally refer to "checking" and "applying" results. In projects where the aim is to seek information, validation normally consists of sharing the results. Usually, validation actions are included in the institutionalisation of what has been learned and implicitly there is a mixture of both concepts. In any case, a *dialectic of validation*, as formulated by Guy Brousseau (2002), cannot be found in the explanations of the project method or in the published examples. However, the promoters of the method did insist on the need for the *institutionalisation* of what had been learned. We found in the examples frequent allusions to the need to take stock of newly acquired knowledge. The *didactic technique* used for this encourages children, both during and at the end of the process, to write accounts and summaries of the work they have finished. This general technique can be applied in a number of ways, using summaries, reports, collective diaries, notebooks, monographs, etc. Through these devices the teacher tries to institutionalise the work carried out in the project.

The teacher mainly performs management and coordination tasks in collective debates, in composing the aforementioned documents, and in the rest of the process. This is when most of the periods of institutionalisation—and, sometimes, validation—take place. Therefore, those who are more critical of or less naïve about this teaching system consider these collective productions indispensable.

4 Contemporaneous theoretical and critical views

The historical period and the institutional conditions under which an educational fact is studied is part of the *ecological analysis* that complements the analysis of didactic practices. As far as the project method is concerned, it seems that, in the context of a crisis in educational systems, new ideas were easily introduced, at least in the sectors most prone to renewal. Those fresh ideas in education were based on philosophical concepts. John Dewey inspired the project method, while W. H. Kilpatrick, professor of education in Teachers College at Columbia University, developed it, although in fact the project method cannot be attributed to just one person, even if several pedagogues, including Dewey, claimed they were the sole authors. This and the lack of a theoretical framework (both conceptual and pedagogical) resulted in the vagueness of the project method (Del Pozo, 2009).

For Dewey, schools needed changes not only in the subjects taught in the curriculum but also in the way of teaching and the way of studying those subjects (Sáinz, 1928/1961). The idea was to fight against intellectualist teaching, which used the textbook as almost the ultimate *source* of information, or the only place to look it up. The only established *method* for teaching was memorising and verbalising the content of books. These innovators could see a need to change not just the goals of the school but also the means. Active methods, and in particular the project method, are a response to this need, which appeared as not only a pedagogical need but also a social need.

What were the ecological conditions that favoured the dissemination of this method? Apart from the new pedagogical ideas, we must consider the efforts of the official institutions during the Second Spanish Republic to renew schools. This included renewing teacher training and creating an environment conducive to experimentation with new proposals. Another factor that contributed to the innovation was the lack of formal programmes and evaluations in primary school. The absence of institutional and social pressure on students could have encouraged the implementation of projects, in which the results of the collective work were displayed to *society* (normally the rest of the school and the parents). These results could be theatrical plays, school newspapers, exhibitions with monographs or collective books, or created materials.

The theoretical principles on which the project method is based—the importance of action as a base for learning and the social character of education—are not exclusive. Actually, they are shared by many of the so-called active methods that were developed and disseminated during that period, such as those of Ovide Decroly (1871–1932), or the Dalton Plan of Helen Parkhurst (1886–1973) or the pedagogy of Carleton Wolsey Washburne (1889–1968) in the Winnetka schools. Margarita Comas and other proponents of the project method include these other methods in their writings aimed at reforming old teaching methods, in which

> laws, principles, definitions and effects have been given a priori instead of being induced; the work has been presented in a series of disconnected rooms

without it being presided over by a unity of objectives and action; the student has not been given a chance or requested to take part in considerations regarding their work, whether it was good, useful or effective [...]

(Sáinz, 1928/1961, p. 33, our translation)

For the two theoretical principles mentioned above, the activities proposed must have two properties: be problematic and develop in their natural atmosphere.

As a refreshing approach and an alternative to traditional methodologies, projects are the main feature of many of the articles, such as the ones cited above, which highlight their advantages and contributions. Nevertheless, other voices drew attention to weak points of the system or simply problems in their implementation. Even Sáinz (1928/1961), who was committed to the method, listed some criticisms in his book and tried to resolve them. Some critics charged that projects lack systematisation and logical rigour. Indeed, the existence of institutionalisation at certain points of the project does not guarantee that all members of the group will acquire an organised body of knowledge. Another of the criticisms that Sáinz commented on was the possibility of mixing topics in the curriculum in a chaotic way, as they are not differentiated. Regarding the objection that projects alter the sequential organisation of teaching, he reminds us that the teacher has the option of formulating a project and deciding how it will be approached so that it matches the question to be dealt with. In a related criticism, he also considered the objection that projects are excessively long.

Other critics, such as Bayón and Ledesma (1934), who had implemented the method in school, also pointed out difficulties and limitations of the method; some of their objections qualified the opinions of those who were considered "experts". In particular they saw the emphasis on spontaneity in the project method as naïvely identifying the school with life outside school. It was considered that school activity should be intentional and that attempts to organise all activities through projects would therefore go astray. Ultimately, the issue concerns the conception of school and the *function of the school* in society:

> We believe that when it is said that school must be a reflection of life, this is only a half-truth. [...] School will start being an image of life; better still, life itself; but a life with aspirations of being better than the life that surrounds us; it must be a constant effort to improve the tone of our existence. If it were not like that, it would not deserve to be called a school.
>
> *(Bayón & Ledesma, 1934, pp. 75–76, our translation)*

Finally, there was a point of agreement between main supporters of the method and people who, while acknowledging its advantages, questioned the way the principles were interpreted as well as some aspects of implementation. This point of agreement was the need for real trial runs, taking into account the *institutional conditions* that accompanied their implementation: one-room classes; high students-teacher ratios; schools separated by gender; etc. An important constraint was the lack of resources—in particular, the lack of "media" (Chevallard, 2006) that students could use for independent research to develop answers to questions asked and questions arising from the research process. Critics pointed out other difficulties, derived from the usual pedagogical model, such as the limitation that children were not used to working in teams. Some drawbacks resulted from the method itself, while others were related to the institutional constraints of the Spanish schools of that period.

Tragically, the Spanish Civil War and Franco's dictatorship abruptly interrupted any renewal process in teaching, eliminating the chance to perform analyses with a certain perspective. No additional indicators were therefore developed for its evaluation.

5 The project method and mathematics

We will now consider the contributions and limitations of the project method when applied to learning mathematics, considering the characteristics of the method and the theoretical approaches described, specifically the elements of the pedagogical ideology of reference, together with the epistemological model of mathematics built from the examples compiled in published works. To study the ecology of the project method as a didactic proposal, we consider the constraints existing at the first *levels of didactic codeterminacy* (see Glossary)—that is: a) the levels of *society* and *school*; b) those resulting from the dominant *pedagogy* in institutions related to primary school teaching, in particular the specific and prevailing way to interpret *learning* and *teaching mathematics*; c) those derived from the *epistemological model of mathematics* that prevailed in those institutions. Our analysis seeks, on the one hand, to better understand the contributions of the PM in teaching and, on the other, to determine the causes of troubles or obstacles that might arise in connection with mathematics, some of which have been perceived by proponents of the method.

Some characteristics of the predominant school practices up to that time were memorisation and verbalisation, with individualistic teaching based mainly on the use of the textbook. By contrast, the PM uses new pedagogical ideas, such as the view that the child should be at the centre of New School processes—ideas that were only new in terms of their implementation in the classroom. The goal of this new *didactic device*, with its corresponding *study gestures*, was to provide children with more functional learning opportunities while taking into account the conditions of the teaching system in schools.

Before the PM, children studied subject matter without knowing why it was chosen. This corresponds to what Chevallard (2006) called a "monumentalist" pedagogy, displaying knowledge as finished "works" (of mathematics) to be "visited" without any motivation to study them and without knowing why they exist. By contrast, projects must interest a child, and often they emerge from or are suggested by the teacher as a result of real situations. With the PM, disciplines are not studied separately but rather appear related through problems that, arising from life itself, are not limited to the sphere of a single discipline. Analogously, the documentation sources in the PM are not restricted to a single textbook but include *media* such as the press, library resources, reference books, interviews with professionals or experts. Furthermore, the PM offers a collective process of study, in which the class works as a real *study community*, in the sense used in the ATD.

The above suggests that the PM prompted a change in the dominant pedagogy of the time. Nevertheless, if a new *pedagogical model of reference* seems to exist, we do not observe a similar change in the way mathematics is interpreted, which would point to an evolution towards a new *epistemological model of reference* (see Glossary) in mathematics. The defenders of projects admit a certain specificity to instrumental subjects—reading, writing and arithmetic—but do not consider it indispensable to design projects to work on these subjects, arguing that they are present in all projects that are carried out:

> [A]s language (reading and writing), as well as arithmetic, are included in everything, […] it follows that there will be four subjects in the program that will take part in the projects, even if for the moment there is no specific one for arithmetic or grammar.
>
> *(Comas, 1931/1963, p. 14, our translation)*

When new knowledge or techniques are needed that have not been taught previously, they are explained then and there, either by stopping work or by deploying what Comas called "accompanying projects" or "included partial projects" (Bayón criticised these auxiliary studies when their extension branched off too far from the main project). For example, when the proposed project was to calculate the capacity of the reservoir of a fountain (the basin), children were presented with the following statement: "Calculate the litres of water that fit in the reservoirs you have in the orchard. They are [rectangular] parallelepipeds. You will take the measurements. You have to measure length, width, and depth" (Bayón & Ledesma, 1934, p. 54). Later, the teacher specified: "I have written on the blackboard these formulae [to calculate the circumference, the area of a circle, and the volume of a cylinder] to help them" (p. 67).

In the project that consisted of making wooden balls for a foundry that would manufacture them in iron, commented on by Sáinz (1928/1961), "the geometry class *said how* they would find the volume of the spheres" (p. 46, our translation and italics). There are no indications that this was done using a method much different from the traditional one.

Another feature of projects is that, when organising the study of mathematics, they do not necessarily consider certain constraints related to the nature of this science. Regarding the function of mathematical knowledge, there are no references to the function of projects in the specific case of teaching and learning mathematics. It would be essential to choose and make students face a "suitable" problem, that would not only give meaning to a new mathematical notion but also induce the building of answers, at first partial and provisional but that would finally lead to the genesis or introduction of such notion. The issue is not so much that mathematics appears primarily in the projects as auxiliary tools for other sciences, but rather that they only appear in this way. When we wonder about the underlying epistemology of this method, we find some attributes that belong to what has been called "applicationism" (Barquero, Bosch & Gascón, 2013): a separation between the social or experimental sciences and mathematics. These authors imply that mathematics has the function of providing the former with a quantifying tool for the study of social or scientific phenomena. We commented above that, in some cases, mathematical work appears separate from the main projects, and constitutes in itself a supplementary project, often dispensable. Therefore, this method promotes a mathematics focused on applications, in contrast to the construction of mathematical knowledge.

> *When interest is lost in the project,* it can be considered finished, and other interested parties (parents, friends, etc.) invited to come and see it. The distribution of invitations and preparation of the last few details *will motivate arithmetic exercises* (calculating how many cards are needed, according to the ones each child requests, cut the ones that have already been made, pile them in tens, see how many we can make from each piece of paper, deduce how many pieces will be necessary, etc.)
>
> *(Comas, 1931/1963, p. 28, our translation and italics)*

Projects may stimulate the acquisition of mathematical knowledge, but they are not planned as a means to gain access to that knowledge. They simply provide them with a context that gives them a raison d'être.

The predominance of that instrumental character over its formative character affects the status that mathematical notions studied may attain, in the sense that it favours the function of mathematical objects as *tools* (instruments to solve a problem), instead of their consideration as *objects* of study, set in the construction of an organised body of knowledge (Douady, 1986).

These deficiencies in the method were pointed out by José María Eyaralar Almazán (1890–1944), a teacher trainer who made some interesting proposals at that time concerning the teaching of mathematics (Sánchez-Jiménez, 2015). In his *Metodología de la Matemática* (Eyaralar, 1933), he doubted the alleged advantages of the PM when the intention is to teach arithmetic exclusively through this method:

> [The project method] can be considered as an important part of teaching, even the basis of it, but it is no substitute for the systematic teaching of arithmetic [...], or the systematic part of our science, which provides the greatest part of its educational value, unless, between the *projects* that we can consider, we could include mastering arithmetic or systematising knowledge.
>
> *(p. 256, our translation)*

It is interesting to see that, in spite of the differences in the political, economic and social context, some of the criticisms, difficulties and constraints associated with the PM in the Second Spanish Republic do not seem to differ a lot from what can be found in today's teaching proposals based on the paradigm of questioning the world. This will allow us to make some of the analytical and design tools provided by the ATD more precise and illustrate their differences with other inquiry-based teaching proposals, such as the PM.

6 The project method and the ATD: the notion of study and research path

We shall first contrast the project method with the notion of *study and research path* (SRP) as it is worked through in the ATD (Chevallard, 2004): not just as a model to devise, design, and describe study processes, but also as a reference for their analysis. An SRP arises from a key question, known as the "generating question", whose study can generate several mathematical organisations allowing us to answer the questions that have motivated the construction of these organisations, all in a functional way, with mathematical modelling work underlying the whole process. We shall now examine which features of SRPs are shared with projects and which are not (Sánchez-Jiménez, 2015, pp. 409–453).

In both, "real" problems are studied, i.e. problems arising from the students' environment, close to their areas of interest, and fostering intrinsic motivation. However, regarding mathematics, the projects themselves do not usually include *generating questions* in the SRP sense; that is, there are no questions or problems that, when solved, lead to the construction of a mathematical organisation. However, with other disciplines, mostly the disciplines that underlie the project's subject theme, a question with these characteristics could in some way be considered to exist.

Teaching within the framework of an SRP is not conceived as a process to be organised individually—concepts are generated in a *study community*. In the same way, with the PM, the individualistic conception of study has been superseded by a designed process that includes individual work, teamwork and class work.

One feature of SRPs is that, at the time of formulating a technique, even though teachers have a guidance role, they do not provide the final technique. However, in contrast to other disciplines, in the case of mathematics it is not clear from examples how the PM guarantees the creation and mastery of mathematical techniques. Moreover, these techniques are handled "outside" the project, by children not taking part in it. Consequently, the role of the teacher as director of the study process, which is much clearer when working with projects in the humanities or experimental sciences, does not remain the same when mathematical questions arise.

The structure of SRPs is tree-shaped, as the initial question leads to others, whose answers help produce a response to the question posed. In projects too, questions arise, prompted by the work underway, but these are not always key factors when building a response to the initial question. This relative independence from mathematical questions studied in projects hinders the development of another feature of SRPs, which is to make it possible for the mathematical organisations built to be integrated within more complex and general ones.

In an SRP, there must be measures for testing and validating answers, not only at the end but also during the study process. According to Dewey, the project method was developed so that the student can "plan solutions that one must be responsible for, and develop them methodically", and so that "one has the opportunity and chance to try ideas, applying them in a way that makes their meaning clear, and discovering their validity by oneself" (*La Escuela Moderna*, 1934, p. 236). However, for some mathematical questions, the teacher provides the answers in the form of procedures or techniques.

Finally, both the SRP and the projects go beyond the sphere of the classroom. Even projects that are framed in a specific subject (e.g. mathematics) are not circumscribed by that sphere. And over the school year the set of SRPs, like projects, must cover the full curriculum, even though there might be redundancies, which are considered necessary, unavoidable and even useful.

7 The project method and the ATD: the notion of inquiry

We shall now examine the project method from the wider and more formal perspective opened up by the general notion of *inquiry* as understood in the ATD. We trust that our readers will be broadminded enough to tolerate the use of symbolic formulas to supplement ordinary language. Contrary to most uses of the word in the literature on education and educational research, in the anthropological theory of the didactic the notion of *inquiry* is used as a modelling tool in studying the didactic. It is therefore not used restrictively to refer to specific pedagogic modes, as is often the case, for example, in the literature on "inquiry-based learning". Let it be known also that an *instance u* is either a *person x* or an *institutional position p* in an institution *I*. We shall say that an "observing instance" *w* identifies an emergent system of inquiry involving the instance *u* whenever the observer *w* regards *u* as making a gesture consisting in considering as problematic a supposed question Q. This inquiry situation can be denoted thus: $w \vdash S_0(u, Q)$, which can be read as "according to the observing instance *w*, the instance *u* considers the question Q as problematic". The observing instance *w* can be a didactic research instance, and in particular a didactician ξ, in which case we shall write $\xi \vdash S_0(u, Q)$. Under what conditions such inquiry situations occur and what processes will ensue are the fundamental problems of the *theory of inquiry* as a subtheory of the ATD.

Let us give some simple but important examples. By far the most frequent case seems to be the following: it appears to the observing instance *w* that the instance *u* considers the question Q and, almost instantaneously, consigns it to oblivion and indifference, almost as if the instance *u* had never encountered Q. From the point of view of the observing instance *w*, the inquiry initiated by the instance *u* is thus "stillborn". This, it seems, is the most common fate of questions in many an institution: they die as soon as they come to life. The second type of inquiry situations involves an instance $\bar{u}$ to whom—from the point of view of the instance w—the instance u puts the supposed question Q, a fact we write as $w \vdash S_1(u, Q, \bar{u})$, which can be read as "according to the observing instance w, the instance u entrusts the study of the question Q to the instance $\bar{u}$". We can have $\bar{u} = u$, in which case u puts the question Q to itself, or even $\bar{u} = w$ (and, in particular, $\bar{u} = \xi$). Subtypes that may easily come to mind are the following: u is a teacher and $\bar{u}$ is one of u's students; or u is a teacher of a school class $\bar{C}$ and $\bar{u}$ is the student position in $\bar{C}$; or, again, u is some instance that can (legitimately) ask the class $\bar{C}$ (teachers and students) to inquire into the question Q. Another subtype may come to mind less spontaneously. In this case, u is the same as before—it can be an agency that (legally and politically) decides on the class $\bar{C}$'s curriculum. But the instance $\bar{u}$ is now some teacher's position, teachers in that position being required by u to inquire into some question Q—for example, "What is the derivative of a function at a point and what should a teacher teach their students about it?" Note that such questions often remain implicit, the instance u being satisfied by tersely stating in this case: "Derivative of a function at a point."

This example should remind us that the visible activity of a class $\bar{C}$ is the forefront of a bigger system, the background of which largely consists of inquiry work done by teachers on their own—in their own *topos*. In all cases, the inquiry system S becomes $S_2(\bar{u}, Q)$. We can sum up all this by the following sequence of systems: $S_0(u, Q) \rightsquigarrow S_1(u, Q, \bar{u}) \rightsquigarrow S_2(\bar{u}, Q)$, in which the "wave" arrow $\rightsquigarrow$ means "gives way to": the instance u considers Q as problematic, then entrusts the study of Q to the instance $\bar{u}$, and finally the instance $\bar{u}$ inquires about Q. The case when $\bar{u}$ is a school class $\bar{C}$ provides a paradigmatic example of what often happens. The positional instance $\bar{u}$ (i.e. the position of being a member of $\bar{C}$) then splits up into two subpositions, $\bar{u}_s$ ("student") and $\bar{u}_t$ ("teacher"). The subjects of $\bar{u}_s$ are supposed to go on the hunt for an answer A to the question Q while those in position $\bar{u}_t$ are, in some form or other, supposed to supervise the students in their search for A. We therefore arrive at a *study and research system* $S_3(\bar{u}_s, \bar{u}_t, Q)$, usually written $S(X, Y, Q)$, where X and Y are the sets of persons occupying the positions $\bar{u}_s$ and $\bar{u}_t$, respectively, in $\bar{C}$.

When it starts to operate, the S&R system $S(X, Y, Q)$ is seen to gradually build up an S&R *milieu* M and a provisional answer $A^\heartsuit$, this process being schematised as follows: $[S(X, Y, Q) \rightarrowtail M] \hookrightarrow A^\heartsuit$. We can make this schema more precise by introducing the *time* variable (either discrete or continuous). At the instant $\mathfrak{t}$, we have: $[S(X, Y, Q) \rightarrowtail M_t] \hookrightarrow A^\heartsuit_t$. It is supposed that X and Y, and of course Q, remain unchanged during the S&R period T. If M and $A^\heartsuit$ are the final milieu and answer, we thus have:

$$S_0(u, Q) \rightsquigarrow S_1(u, Q, \bar{u}) \rightsquigarrow S_2(\bar{u}, Q) \rightsquigarrow S_3(\bar{u}_s, \bar{u}_t, Q) = S(X, Y, Q) \rightsquigarrow \{[S(X, Y, Q) \rightarrowtail M_0] \hookrightarrow A^\heartsuit_0\}$$
$$\rightsquigarrow \dots \rightsquigarrow \{[S(X, Y, Q) \rightarrowtail M_t] \hookrightarrow A^\heartsuit_t\} \rightsquigarrow \dots \rightsquigarrow [S(X, Y, Q) \rightarrowtail M] \hookrightarrow A^\heartsuit.$$

The set M_f comprises partial answers to Q, new questions and tentative answers to these questions, and works of all nature gathered to use them as tools in the S&R process.

The general model sketched here is meant to apply to *all* situations in which a question is raised or even an answer to an unstated question is implicitly given—like the "Derivative of a function at a point." In particular it is postulated that *all* school class situations, whether innovative or traditional, are analysable in terms of the concept of inquiry, understood as expounded here. The formal schema above suggests a host of questions, notably about the milieu formation process and the path, i.e. the study and research path, taken to arrive at the answer $A^\heartsuit$. Among the questions worth asking, let us briefly consider the following two cases. The first one is: "By whom, and according to what criteria, is an answer $A^\heartsuit$ deemed admissible and relevant?" This is indeed a crucial question from a didactic point of view. We shall denote by the letter v the "evaluating instance" that appraises the answer $A^\heartsuit$. We can have $v = w$, or $v = u$, or $v = \bar{u}$, or whatever. The evaluating instance v judges $A^\heartsuit$ according to a set $\check{C}$ of criteria. One important kind of criteria—that the evaluating instance v may acknowledge or not—is expressed in the following question schema: "Does the answer $A^\heartsuit$ and its validation take into account what the discipline D can say about the question Q and, ultimately, about the answer $A^\heartsuit$?" Here, the discipline D is understood to be *any* field of human experience together with the rules that define it at a given time (its "discipline"). When a question Q is considered by some instance u within an institution I, there always exists a "cultural distance" (depending on the culture of the institution I) between the question Q and the given field D, which may lead the instance u to hold, often without realising it, D as irrelevant or little relevant to the study of the question Q. This is particularly true when D is the field of mathematics, at least whenever the institution I is immersed in a mainly literary culture, as is often the case. For lack of space, we shall not dwell more, here, on this essential point of *any* inquiry, both theoretically and practically.

The second announced question is: "Where does the 'generating' question Q come from?" The most general response provided by the ATD is the following: a question Q stems from some project Π. A project is simply defined as a *task* of some kind that some instance u intends to perform. Now this task may have problematic aspects for u, which leads u to consider some question Q as an open question. Let us note here that, usually, a question Q ensues from *several* projects Π. The instance u may then seek help from an instance $\bar{u}$, so that we will observe the forming of a system $S1(u, Q, \bar{u})$, in which u is the "demander" and $\bar{u}$ is the "demanded", the demand being an appropriate answer $A^\heartsuit$ to the question Q—which will eventually have to be implemented by u. Of course, we can have $\bar{u} = u$: this is even, it seems, the central premise of the project method as originally designed. We know from the above study that this creates obstacles that are difficult or practically impossible to overcome. One main stumbling block is the project Π itself, which, from the point of view of educational decision-makers, may consume much too much time and energy on mostly unproblematic tasks, with low or even negative educational returns. This does not lead to discarding the project method but to distinguish more realistically between the instances we have denoted by u and $\bar{u}$, who can both continue to fully be classroom instances, the implementation of projects Π (few and far between) and the inquiry into questions Q (the marrow of daily life in the classroom) going at appropriately different paces. More generally, the modelling in the framework of the ATD of empirically observed or merely imagined modes of study can take advantage of a number of key parameters—such as the student's and teacher's *topos*, for example—that are casually merged together or remain implicit in ordinary approaches to the didactic.

8 Final thoughts

The analysis conducted here portrays the process of studying mathematics within the framework of the PM, using, to some extent, tools provided by the ATD. One aspect of the project method, apparently inherited from general pedagogy, which seems to have inspired this teaching method, is that normally, when organising the study of mathematics, projects do not take into account the constraints resulting from the nature of this science. Although the most innovative teachers generally acknowledge the virtues of the method, others, like Eyaralar, have views about the nature of mathematics that make them more critical of the PM.

In this respect, the above offers an example of what happens when innovations in the pedagogical and teaching model are not accompanied by changes in the epistemological model that, more or less consciously, is being relied upon. The consequence is that pedagogical innovations in environments prone to renewal may sometimes lack interaction with real innovations in the field of knowledge at issue—perhaps most notably in mathematics.

Today, projects as an approach have been revived in the teaching world, with voices attributing to them properties quite similar to those described in the period studied here. There does not appear to be currently any rigorous questioning to determine what mathematical knowledge can be learnt through projects under appropriate design and implementation conditions. This is why research on the design and implementation of SRPs driven by the concept of inquiry is key.

References

Barquero, B., Bosch, M., & Gascón, J. (2013). The ecological dimension in the teaching of mathematics at university level. *Recherches en Didactique des Mathématiques*, 13(3), 307–338.

Bayón, D., & Ledesma, A. (1934). *El método de proyectos: Realizaciones*. Madrid: Escuelas de España.

Brousseau, G. (2002). *Theory of didactical situations in mathematics*. Dordrecht, The Netherlands: Kluwer.

Chevallard, Y. (2004). Vers une didactique de la codisciplinarité: Notes sur une nouvelle épistémologie scolaire. Journées de didactique comparée 2004, Lyon, France.

Chevallard, Y. (2006). Steps towards a new epistemology in mathematics education. In M. Bosch (ed.), *Proceedings of the IV Congress of the European Society for Research in Mathematics Education* (pp. 21–30). Barcelona, Spain: FUNDEMI-IQS.

Comas, M. (1931/1963). *El método de proyectos en las escuelas urbanas.* Buenos Aires, Argentina: Losada. Del Pozo, M. M. (2003–2004). La escuela nueva en España: Crónica y semblanza de un mito. *Historia de la Educación,* 22/23, 317–346.

Del Pozo, M. M. (2009). The transnational and national dimensions of pedagogical ideas: The case of the project method, 1918–1939. *Paedagogica Historica: International Journal of the History of Education,* 45(4/5),453–693.

Douady, R. (1986). Jeux de cadres et dialectique outil-objet. *Recherches en Didactique des Mathématiques,* 7(2), 5–31.

Eyaralar, J. M. (1933). *Metodología de la matemática.* Madrid: Reus.

Knoll, M. (2014). Project method. In D. C. Phillips (ed.), *Encyclopedia of educational theory and philosophy* (vol. 2, pp. 665–669). Thousand Oaks, CA: Sage.

La Escuela Moderna (1933). El método de proyectos en la enseñanza (continuación). *La Escuela Moderna,* 507, 562–569.

La Escuela Moderna (1934). El método de proyectos en la enseñanza (continuación). 510, 124–134 & 512, 230–240.

Mata, A. R. (1923). Un nuevo método de enseñanza: El project method. *Revista de Pedagogía,* 18, 206–211.

Sáinz, F. (1928/1961). *El método de proyectos.* Buenos Aires: Losada.

Sánchez-Jiménez, E. (2015). Las escuelas normales y la renovación de la enseñanza de las matemáticas (1909–1936) (Doctoral dissertation). Universidad de Murcia, Spain.

Stevenson, J. A. (1921). *The project method of teaching.* New York: The Macmillan Company.

Stimson, R. W. (1954). Home project teaching and related educational developments. In R. W. Stimson & F. W. Lathrop (eds), *History of agricultural education of less than college grade in the United States* (pp. 582–606). Washington: US Office of Education.

Viñao, A. (1994/95). La modernización pedagógica española a través de la "Revista de Pedagogía" (1922–1936). *Anales de Pedagogía,* 12/13, 7–45.

8

THE ECOLOGY OF STUDY AND RESEARCH PATHS IN UPPER SECONDARY SCHOOL

The cases of Denmark and Japan

Britta Jessen, Koji Otaki, Takeshi Miyakawa, Hiroaki Hamanaka, Tatsuya Mizoguchi, Yusuke Shinno and Carl Winsløw

1 Introduction to our research question

A fundamental challenge for mathematics education, presumably in a very broad sense that would encompass most levels and institutions worldwide, was pointed out by Yves Chevallard (2015, p. 177) in his regular lecture for ICME 12:

> The relation to knowledge and ignorance [...] associated with the visiting of mathematical works has become increasingly unsuited to people's needs and wants, up to the point that there currently exists a widespread belief that mathematical knowledge is something one can almost altogether dispense with ... .

The "visiting of mathematical works" refers to a more traditional way of studying prescribed mathematical topics, such as quadratic equations or the addition formulae, which constitute the core in remarkably similar mathematics curricula across the world. Frequent tests and exams certainly supply enough institutional pressure to ensure that many students succeed in acquiring, at least passingly, some of the mathematical procedures and ideas that the school presents them with—but even if they recall some of it afterwards, the mathematical needs they then face will rarely resemble routine school exercises. A more *autonomous* relationship to knowledge is needed (ibid., p. 178):

> a receptive attitude towards yet unanswered questions and unsolved problems, which is normally the scientist's attitude in his field of research and should become the citizen's in every domain of activity.

To develop this attitude or relationship towards knowledge in general, and to mathematical knowledge in particular, Chevallard proposed an innovative format for school teaching, the so-called *study and research paths* (SRPs), in which students, guided by a teacher, engage in work, not with monuments or "answers", but *questions*. In short, an SRP is a didactic process initiated by a more or less challenging *generating question*, pursued through the production of *derived questions* and the search for *answers* to these questions in a dialectic of *study* (roughly, search for relevant information) and *research* (interaction with a milieu in the sense of the theory of didactic situations (Brousseau, 2002)).

Generating questions should be of importance to society and its citizens, and it goes without saying that students at a given age level may be able to study only some of these primary questions. In this set-up, the consultations of "works" (texts and other media) will not disappear, but it will regain its rationality as means to the study of a question. The main point of this proposal is the aim to educate new generations to develop a relationship to works as *resources* for inquiry, which one needs to consider critically in terms of validity and pertinence; the study of resources is motivated by a question with which one is confronted, and for which no predefined or simple answers are available.

This vision of education in general—and in particular, of mathematics education with all its manifest challenges—is likely to appear attractive to many readers, and it certainly resonates, with both current trends such as "inquiry-based mathematics education" (Artigue & Blomhøj, 2013) and more classical paradigms like "problem solving" (Schoenfeld, 1985) and "critical mathematics education" (Skovsmose & Nielsen, 1996). The main novelty of SRPs, to us, is the combination of three strongly related elements: the audacity of the didactic vision (outlined above), the strong epistemological theory behind the design tool (Winsløw, Matheron, & Mercier, 2013), and an international and empirical research programme which works with it. SRPs are thus a concrete proposal for "questioning the world". The research questions we pursue in this chapter are:

- What conditions and constraints will SRPs currently meet at upper secondary level within (very) different societies like Denmark and Japan[1]?
- What further conditions are needed to make SRP viable in these institutions?

Our inquiry is supported by the didactic, pedagogic and societal context found in these countries, including a selection of documents and other sources that can provide partial answers to our questions, along with results, observations and anticipations related to the preliminary experimentations, which have been carried out there.

2 Theoretical framework: variables considered

In Denmark and Japan, only a few and small-scale experiences with SRPs exist so far (e.g. Jessen, 2017a; Otaki, Miyakawa, & Hamanaka, 2016). Our main purpose is not to report on these, but rather to analyse their ecology, which means to analyse and elucidate the conditions that support the realisation of SRPs in these countries and the constraints that hinder it. These will be situated at different levels of the scale of didactic co-determinacy defined by Chevallard (2002) (Figure 8.1, see Glossary).

It is certainly conceivable that the inquiry on some generating question (e.g. "If the children are only allowed to withdraw their money when the amount is doubled, how long should they wait?") (Jessen, 2017a) does not exceed the boundaries of the disciplines. Yet, as we shall see, many questions that are currently favoured by conditions at higher levels of didactic codeterminacy (e.g. *what causes financial crises?*) will rapidly lead to, or even require, an inquiry into questions that call on resources and methods from different school disciplines.

Our analysis below, of constraints and conditions for introducing SRP-based teaching in our school systems, is guided by a number of parameters. Bearing in mind the scale of levels of didactic codeterminacy, we focus on the ecology of SRPs situated at the levels from school to discipline, and in particular on two entities that are influential for didactic situations at school: *school curricula* and the conditions for the *teaching profession*.

As SRP can be regarded as an approach to inquiry-based pedagogy "in which students are invited to work in ways similar to how mathematicians and scientists work"

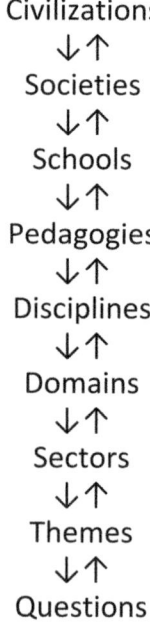

FIGURE 8.1 The scale of levels of didactic codeterminacy

(Artigue & Blomhøj, 2013, p. 797), we need to know why and how societal and civilisational conditions affect curricula in our countries, in terms of inquiry-based teaching. We also investigate the extent to which curricula support applications of mathematics and cross-disciplinary activities. As argued above, the nature of the generating questions of SRPs leads to very different possible paths—i.e. derived questions and answers—from purely mathematical ones to paths drawing on other disciplines.

A major restriction on teachers' work in the classroom, particularly at the secondary level, comes from high-stake exams and their backwash on teaching (Schoenfeld, 1988). Therefore, we consider how forms of assessment align with inquiry-based mathematics education and in particular the SRP format. In addition, different kinds of *resources* and *systems* for teachers' professional education and development can affect the use of SRPs and other forms of inquiry-based mathematics teaching (e.g. Dorier & García, 2013). More precisely, we consider, for each of the two societies, how the design, implementation and continued use of SRPs may be influenced by current *didactic* infrastructures—the set of conditions affecting teachers' work in the classroom (Chevallard, 2009)—and *paradidactic infrastructures*—systems and resources for teachers' professional development, including initial teacher education (Winsløw, 2011; & Winsløw et al., 2013).

This leads to the following questions, which represent six variables for our analysis:

1.   How are ministerial/official guidelines and the curriculum developed?
2.   Do the ministerial guidelines or the curriculum support inquiry-based mathematics education? For instance, do they include formulations like "application of mathematics" or do they support cross-disciplinary activities based on the study of questions that are of importance to them and society?
3.   How are textbooks and teaching materials developed? What legitimacy do they have in relation to teaching practice?
4.   How is the teaching and learning of mathematics assessed, including what kind of assessment regulates the entrance to higher education and how might this affect the implementation of SRP-based teaching?
5.   To what extent do paradidactic infrastructures favouring the implementation of a SRP teaching approach exist within the school system? To what extent does teachers' initial education support teachers' own development of inquiry-based teaching?
6.   Is the daily use of ICT and the internet as a means to construct new knowledge an integrated part of the school system?

We now proceed to analyse these variables for our two contexts, to identify the potentials and obstacles for SRPs to become viable.

3 Conditions and constraints found in Denmark

The subsections below and in the next section correspond to the six variable above.

3.1 The official regulation of mathematics education in Denmark

In Denmark, the Ministry of Education formulates the curriculum and other regulations for upper secondary schools and organises the centralised high-stake written exams. There is a tradition of broad political agreements concerning the purposes, goals and regulation of schooling in general. For instance, the general frame for the teaching of mathematics, English, and cross-disciplinary elements (Danish Ministry of Education, 2016a) was formulated based on a political agreement made between most of the parties in the Danish parliament. Thus, through elected politicians, society affects teaching through regulations.

Based on such agreements, the Ministry of Education formulates further detailed regulations, e.g. about the disciplines and exams in upper secondary education, which must be proved by the minister and the involved parties (Danish Ministry of Education, 2017a). In August 2017, a new reform was implemented. The act provides the students with the possibility of choosing between 18 different study lines (ibid., p. 11). A study line is defined by three disciplines at certain levels, e.g. mathematics A, physics B and chemistry B (A being the highest level and C the lowest). Together with these three disciplines, the students have mandatory subjects: Danish, English and history, among others. The act has explicit passages on mathematics: for instance, students are supposed to be guided in their choice of study line, based on a screening test in mathematics after three months in high school. The act further declares that most students should have mathematics at level B or higher.

In this chapter, we refer to the new curriculum of mathematics B (Danish Ministry of Education, 2017b). The Ministry of Education appoints a group of experienced teachers to formulate the actual curriculum, following several overall regulations and political directives.

The parts of the political directives which specifically concern mathematics are in part inspired by an evaluation report on upper secondary mathematics education (Jessen, Holm, & Winsløw, 2015), which was mandated by the ministry to identify the current challenges for the teaching and learning of mathematics in upper secondary school. Based on the report, a political agreement required the appointment of a mathematics commission, to suggest changes in the curriculum and other regulations of the discipline. The commission provided a report, which has further inspired the new curriculum (Grønbæk *et al.*, 2017). The Mathematics Commission consisted of university teachers, secondary teachers, teacher educators for primary and secondary school, a representative of secondary schools' headmasters, and representatives from private companies. In this way, different quarters of society affect curriculum development.

Both reports (Jessen *et al.* 2015; Grønbæk *et al.*, 2017) suggest that probability theory should again be part of the curriculum. The mathematics commission further suggested that discrete mathematics should be a mandatory topic. Both topics were promoted by researchers and representatives from higher education, affecting what has been included in the new curriculum (Danish Ministry of Education,

2017b). In reality, the documents are compromises between many different political interests and not just the result of inputs from researchers and teachers.

3.2 Inquiry-based mathematics education and curriculum

Both the former and the new curriculum in mathematics support inquiry-based activities. In the previous curriculum, mathematical modelling and how this can be realised (Danish Ministry of Education, 2014) are described, following ideas described by Artigue & Blomhøj (2013). This represents a condition for realising mathematical modelling through SRPs. However, the suggested algorithmic nature of modelling activities does not really reflect the idea of an SRP and does not properly describe actual modelling outside of the school context (see Jessen, 2017b). The present curriculum also emphasises that students should gain knowledge about how to develop mathematical knowledge in interactions between applications of mathematics and knowledge construction. It is formulated as:

> Through an experimental approach to mathematical topics, problems and exercises must include students' mathematical notions and innovative abilities be developed. This means by e.g. the design of inductive teaching activities where students are offered the opportunity to autonomously formulate hypotheses based on certain examples.
>
> *(Danish Ministry of Education, 2014)*

It is further encouraged that teachers support students' development of mathematical notions by carrying out mathematical experiments assisted by CAS^2. These constraints and conditions provided by the curriculum have been explored in SRP case studies (Jessen, 2014; 2017b).

The 2017 curriculum continues to emphasise modelling—that is, mathematising different situations from extra-mathematical situations (e.g. Blomhøj & Kjeldsen, 2006). The introduction to mathematics as a school discipline includes the formulation:

> When hypotheses and theories are formulated mathematically, new insights are often gained. The widespread application of mathematics and mathematical methods in modelling and problem solving is bound in the potentials of the discipline to capture and describe how many very different phenomena behaves in similar ways.
>
> *(Danish Ministry of Education, 2017b)*

Further, it is stated that: "Concretely students must gain competences to understand, formulate and treat problems about real-world problems, as well as knowledge and skills to exercise mathematical reasoning and logical thinking" (Danish Ministry of Education, 2017b). What can be gained from SRP-based teaching is part of the purpose, and under the headline "didactical principles", we find the same passage as previously regarding inquiry-based student work to formulate questions and hypotheses. Furthermore, the new curriculum explicitly emphasises that students must learn to read mathematical texts—Jessen *et al.* (2015) documented how one third of the teachers did not require or expect the students to read the textbook.

Upper secondary students have to select a so-called "study line". Most of these include mathematics as a defining discipline, and in all study lines, the teaching should seek to emphasise connections with and integration of the defining disciplines. In combination with social science, a typical topic is Chi-squared test when dealing with opinion polls. In biology, functional models of population growth are common, as are probability models of genetic phenomena. In physics, classical mechanics is combined with calculus, and in chemistry the teaching of reaction kinetics is based on elements of differential equations. Many of these topics could host the initial question of a modelling activity but, as discussed in Jessen (2017b), many of these questions are, in reality, reduced to a story from real life "hiding" a mathematics exercise, giving students only a limited idea of how, when and why mathematics applies to other disciplines. Examples from textbooks are in general close to "applicationism" in the sense of Berta Barquero, Marianna Bosch and Josep Gascón (2013), presenting idealised non-mathematical situations where mathematics can be directly "applied", with almost no modification. Anyway, both the general organisation in study lines, and the cross-disciplinary "study line project" element could favour the study of generating questions. The potential in relation to SRPs has been experimented with in projects combining biology and mathematics by Britta Jessen (2014), who concludes that the SRP format provides a strong tool for designing and monitoring students' integrated and autonomous work with two disciplines, under existing conditions.

3.3 Textbooks and teaching materials

In Denmark, textbooks are written and developed by experienced teachers. Some textbook authors are part of the groups that formulate the curriculum and the written exam exercises. Therefore, we can presume a reasonable alignment between the intended curriculum (meaning the ministerial documents) and assessed curriculum (meaning what is tested at the written exam)—and the textbooks. The Danish textbooks cover more than what is tested in the written exams, e.g. ideas for projects, inductive activities and examples of mathematical modelling (e.g. Grøn, Felsager, Bruun & Lyndrup, 2012; Clausen, Schomacker & Tolnø, 2010). We do not have any accounts of the extent to which these parts are used in the teaching. The textbooks are often organised as following: a short introduction or motivating narrative, the introduction of new notions, definitions or theorems, some examples of how to use those in exercises, and the end of the chapter, which provides students with exercises similar to the examples just provided. In many cases, theorems are followed by proofs. Jessen (2017b) argues that even chapters on mathematical modelling tend to be "applicationist" and present hidden mathematics exercises rather than authentic modelling problems. However, the two case studies by Jessen (2014, 2017a) indicate that the textbooks could nevertheless fulfil new fruitful functions for students during study processes involved in SRP-based teaching.

3.4 High-stake exams and assessment

In Danish high school, the disciplines may have both oral and written examinations, which are "high stakes" as they determine students' subsequent options in higher education. The written exam in mathematics runs for four or five hours in (for example) a gym hall. During the first hour, computers are not allowed. After this hour, students hand in the first exercises. The rest of the time there is permission to use computer CAS tools and webpages that the student has accessed previously. Jessen *et al.* (2015) report that the written exam exercises are of certain routine types, and interviews with students and teachers show that this promotes "teaching to test" rather than encouraging students' autonomous and applicable knowledge of the mathematical content. The emphasis on preparing students for the standardised written exam represents a constraint for SRP-based teaching (see Jessen, 2017b). The aforementioned study line project is another high-stake exam that may involve mathematics. Each student chooses two disciplines to be involved in their subject. They then write a 15–20-page report, with some supervision from a teacher; the report is assessed as an exam paper. This format of project and exam is a good foundation for an individual SRP, as discussed in Jessen (2014, 2017b).

The oral exam is based on "questions", which are drawn by the students (as in a lottery) just before the examination, where they have about 30 minutes to present their answer in front of the teacher and an external examiner (after 30 minutes of preparation). The questions students draw will often refer to a notion or theorem, which the students must prove. More open questions can be posed and represent favourable conditions for SRP-based teaching, as shown by Jessen (2017a).

The 2017 reform introduced a change in the oral exam. The second part of the oral exam remains the same, but the first part is now a group examination. Students are supposed to work in groups on large problems for 90 minutes in one room with half the class present. During this time teacher and examiner walk around both observing and talking with groups and individuals about the problems and the students' ideas on how to solve them. A student's performance in the group and their individual exam together form the basis of assessing the student (Danish Ministry of Education, 2017b). This kind of oral exam possibly opens up new favourable conditions for the development of SRP-based teaching.

3.5 The existence of paradidactic infrastructures and initial education

It has previously been argued that to introduce inquiry-based teaching into mathematics, teachers need certain paradidactic infrastructures to support the change in teaching practice, such as opportunities for shared preparation and observation of teaching followed by shared reflections (Winsløw, 2011). Such opportunities are not generally available in Denmark. Most sharing of ideas, teaching materials and practice takes place in informal conversation during lunch hours or similar (Jessen *et al.*, 2015). This represents a constraint for the implementation of SRP-based teaching in Denmark. Jessen (2016) argues that to engage teachers in changes of practice in the direction of SRP-based teaching can be compared to the study of a generating question: to engage in study and research processes around such questions requires certain infrastructures, which currently do not exist for Danish teachers.

In Denmark, teachers' initial education consists of a five-year master's degree, with a major and a minor discipline (e.g. a major in mathematics and a minor in physics); teachers can then teach both disciplines. About 90 per cent of this initial education does not target secondary teaching but consists of regular academic courses in the two disciplines. Since 2006 it has been a requirement that teachers wishing to become mathematics teachers in upper secondary education have ten or more ECTS[3] points in didactics and philosophy of mathematics (Danish Ministry of Higher Education and Science, 2006). However, universities run different courses on didactics, which means they vary in content and do not necessarily have an explicit focus on inquiry-based teaching (e.g. SRPs). When starting their first employment after a master's degree, the candidates must complete a practicum. This means that new teachers are teaching half time. In the other half they attend courses on general learning theory and pedagogy, with 2.5 days devoted to courses on teaching practices in each of the new teacher's disciplines (Danish Ministry of Education, 2014). These days are spent discussing how to teach according to the curriculum and guidelines. There is an emphasis on "best practice" presented by experienced teachers. To sum up, only in rare cases do teachers in Denmark have the prerequisites to engage in and develop SRP-based teaching.

3.6 The use of ICT and the internet as a means to construct new knowledge

In Denmark, there is free use of computers for three-quarters of the written exam time for upper secondary mathematics. The last part is "pencil and paper". Free use means using ICT, including CAS tools, for solving the tasks. Further, students are allowed to revisit homepages they have been using during the teaching, including pages as Khan Academy (Danish Ministry of Education, 2016b). Teachers make extensive use of CAS tools and sometimes webpages in their teaching. However, Jessen *et al.* (2015) report on widespread use or production of pre-produced CAS templates for the most common exam exercises. Even when teachers do not encourage or support the development of such templates, they are produced and shared among students (ibid., p. 13). We still consider it a favourable condition for SRP-based teaching that students are conversant with ICT and how to gain mathematical knowledge from webpages.

4 Conditions and constraints found in Japan

4.1 The official regulation of mathematics education in Japan

Japan has a centralised educational system, which provides common educational opportunities across the whole country. The main component, before tertiary education, consists of three stages: six years of elementary school, three years of lower secondary, and three years of upper secondary; the first two stages are compulsory. Today, almost 100 per cent of students enter upper secondary school, and half of them go on to tertiary level (cf. Baba, Iwasaki, Ueda & Date, 2012).

The Japanese education ministry (abbreviated as MEXT) formulates the national curriculum (called Course of Study in English), which determines the objectives and teaching content at each educational stage. This curriculum is revised about every ten years; the current version[4] was published in 2009 (MEXT, 2009a), and a new one is coming soon. In addition to this national curriculum, which describes objectives and content in a simple manner, the ministry publishes official guidelines (MEXT, 2009b) that explain the meanings of the objectives, details of the content and teaching methods.

A team of officials, mathematics teachers, mathematicians and mathematics educators develop mathematics curricula following the general principles and frameworks created by other committees, which reflect educational innovations and trends, the results of national surveys of student achievement, and so on (cf. Ohara, 2007). Today, the key competencies of PISA, which emphasise the application of mathematics in ordinary life, strongly affect the direction of Japanese educational reforms. This educational trend and others such as the emergence of ESD (education for sustainable development) might support the implementation of SRPs in Japan.

The current upper secondary school mathematics programme includes six courses: Mathematics I, Mathematics II, Mathematics III, Mathematics A, Mathematics B and Application of Mathematics. This division of courses is not made according to mathematical domains, as each course consists of a range of different domains, such as algebra, geometry, statistics, etc. Only Mathematics I is compulsory. The organisation of mathematical content in these courses is principally based on the paradigm of *monumentalism* or *visiting works*: a body of mathematical knowledge is broken down into smaller pieces, which are rearranged and taught one by one like visiting monuments. The exception is the course Applications of Mathematics, which has no specific target mathematical knowledge to teach. We will come back to this course later.

4.2 Inquiry-based mathematics education and curriculum in Japan

In this part we explain a keyword of the Japanese mathematics curriculum, and some didactic organisations intended in the curriculum, from the perspective of inquiry-based mathematics education.

The Japanese mathematics curriculum and its guidelines rarely use the term "inquiry". Instead, the term "mathematical activities" is frequently used. This term is situated at the core of the objectives of high school mathematics; any mathematics learning is expected to be carried out through these "activities." According to the guidelines, mathematical activities are "proactive and purposeful activities related to mathematics learning" (MEXT, 2009b). Roughly, this term is intended to foster proactive, autonomous, purposeful and motivated mathematics learning in the classroom, as opposed to learning under the transmission approach, where students receive pieces of mathematical knowledge provided by the teacher. Both aim to foster students' autonomous and purposeful learning. However, this does not mean that the Japanese curriculum is organised within the paradigm of questioning the world. In fact, mathematical activities in the curriculum are merely a means to teach the pieces of mathematical knowledge students are intended to acquire.

On the one hand, mathematical activities are related to every part of school mathematics. On the other hand, there is also a special place for such activities. The so-called *task-oriented learning* (*kadai gakusyū* in Japanese) is newly incorporated in Mathematics I and Mathematics A in the present Course of Study and is intended to fully realise the mathematical activities mentioned above (ibid.). This learning is dissociated from the general teaching content, unlike mathematical activities, and may promote the realisation of SRPs. However, there some constraints hindering SRPs. Task-oriented learning is supposed to be conducted after teaching mathematical content so that students can apply mathematical knowledge already learnt in other domains. This idea is clearly based on the *applicationism* that has been already mentioned in the section on Denmark. The epistemology of applicationism hinders modelling activities (Barquero *et al.*, 2013), and therefore the realisation of SRPs. In general, the dialectic between mathematical models and (extra-) mathematical systems favours SRPs in mathematics: questions or problems related to extra-mathematical situations require mathematical models—e.g. gambling situations produced the mathematical notion of probability—; in contrast, mathematical solutions clarify the nature of extra-mathematical systems—the notion of probability can describe different behaviours of events. Nonetheless, applicationism considers these two poles as independent. As a result, the dialectical process of mathematical modelling is reduced to "first introduce the mathematical theory, and second apply it to a practical situation", which is not what is supposed in SRPs.

In the current Japanese curriculum, application of mathematics and inquiry-based mathematics education are interrelated. While applicationism could hinder the realisation of SRPs, it could also serve as a condition for SRPs. We can easily find in the Japanese mathematics curriculum the influence of an educational approach emphasising the application of mathematics. Task-oriented learning, mentioned above, is one example; there is also another place where the application is emphasised; it is in the optional Application of Mathematics course. This course consists of two domains: "mathematics and human activities" and "mathematical analysis in social life". What mainly distinguishes this course from others is that there is no mathematical knowledge to be taught and learned in advance. The guidelines propose different themes related to humans and society, including games, puzzles, history of mathematics and data analysis. Since no specific mathematics knowledge is aimed at—that is, there is no monument for students to visit—this course favours the introduction of inquiry activities like SRPs. However, very few students actually take this course because it is not covered in university entrance exams. As a consequence, the ministry has abandoned this course and integrated its pedagogical ideas into other courses in the next national curriculum.

There is one more possible place for SRPs outside of mathematics courses but still related to the application of mathematics. It is the subject called Periods for Multidisciplinary Studies (MEXT, 2013; *Sōgōtekina Gakusyū no Jikan*, our translation), which was implemented in curricula at all school levels with the reform at the beginning of the 21st century. As the name implies, this subject aims at multidisciplinary studies. One may find here the term "inquiry" in its statement of objectives (MEXT, 2009c). In contrast with the mathematics curriculum, the curriculum and guidelines for Periods for Multidisciplinary Studies often use this term and situate it at the core of the subject. The Periods for Multidisciplinary Studies course is in this sense a highly appropriate context in which to introduce SRPs. In this subject, students are expected, like practising scientists, to use tools such as the internet and libraries, to go outside the classroom to gather data, to analyse data by any means available, to present the findings in different forms, etc.

4.3 Mathematics textbooks in Japan

In Japan, private publishers produce mathematics textbooks following the national curricula. Textbooks have to be approved by the government before going on sale. Textbook development is usually carried out by a team that includes mathematicians, mathematics teachers, mathematics educators and publishing professionals. In principle, anyone can publish textbooks; in practice, certain publishers more or less monopolise the field.

Since they are required to have government approval, each mathematics textbook more or less reflects the government's educational intentions. However, only two series of textbooks are published for Application of Mathematics, since only a very small number of students take this course, as mentioned above. And the textbooks give only scant attention to task-oriented learning: for instance, 15 pages out of 255 in a textbook for Mathematics I (Takahashi *et al.*, 2015). One may easily imagine that inquiry activities are rarely implemented in actual mathematics classes.

A textbook is generally organised chapter by chapter according to the content area of the curriculum. Each section of each chapter in most mathematics textbooks (except for Application of Mathematics) is organised in a format that first introduces the theory then follows it with exercises for practice, similar to the Danish textbooks based on applicationism. This order of content may also determine the teaching process in the classroom—mathematics teachers in upper secondary school use textbooks in almost all mathematics classes, as required by law. Thus, the organisation of textbooks and its epistemology may hinder the implementation of SRPs.

4.4 Examinations and assessment in Japan

University entrance examinations are one of the biggest events for upper secondary school students and Japanese society as a whole. The system of entrance examinations is slightly different depending on the type of university, public or private. Let us focus here on the National Centre Test for University Admissions (the "Centre Test" hereafter), which all public universities and many private universities adopt for the first stage of examination. This is an achievement test located at the end of upper secondary school (grade 12), organised by a public institution and universities. The Centre Test includes five subjects: Japanese, Mathematics, Social Studies, Sciences, and Foreign Languages. The mathematics test covers four courses out of the six on offer: Mathematics I, Mathematics II, Mathematics A and Mathematics B; students select some or all of them depending on the requirements of the universities they are applying to. The Centre Test uses *scantron technology*, which requires students to mark numbers as answers to the questions instead of writing an answer freely. Students work on the test with pencil and paper, without the use of any other equipment. There is no oral test and no additional module such as a report. In short, the examination should be optimised as a tool for evaluating the students' achievement within the visiting of monuments paradigm.

4.5 Teacher education system and resource system in Japan

There is no doubt that teachers' education and professional development, including mathematical knowledge and didactic knowledge, strongly affects the realisation of SRPs in the classroom. Even though the didactic infrastructure supports inquiry, teachers may not be able to implement it in mathematics classes without knowing how inquiry is carried out. And this is, in fact, the case in Japan. Most upper secondary mathematics teachers graduate from faculties of sciences where they learned mathematics and took some educational classes and practices for pre-service teachers. While they learn many mathematical theories during their undergraduate studies, they get only very limited opportunities to experience mathematical inquiry. Most university lectures adopt the classical transmission approach to teaching mathematics, and even in postgraduate work, which is not supposed to teach specific knowledge, most mathematics professors adopt a teaching method that involves reading a mathematics textbook section by section in a group, without allowing students to realise the rationale of mathematical objects to learn. Thus, undergraduate mathematics students rarely experience the process of inquiry beginning with a lively question. To what extent does such praxeological equipment allow teachers to organise inquiry activities or prevent them from doing so?

The professional development of mathematics teachers is not only carried out in pre-service teacher training programmes in universities; there are also several opportunities—part of the *paradidactic infrastructure*—for in-service teachers to learn in Japan. Lesson study, a well-known format for collaborative work among teachers, is one such (Miyakawa & Winsløw, 2013); among others, the *resources* available to teachers strongly contribute to their ability to elaborate their own ideas on mathematics teaching. Several educators in Japan have published their own instructional *theories* (in the ATD sense) and/or approaches to teaching mathematics. Some examples with relevance to SRPs are *open approaches* (Becker & Shimada, 1997; Nohda, 1991; cf. Miyakawa & Winsløw, 2009;), *problem solving-oriented lesson structure* (Souma, 1997; cf. Asami-Johansson, 2015), *mathematics learning incorporating "question" as a main axis* (Okamoto & Morozumi, 2008), and *researcher-like activity* (Ichikawa, 1998). Each of these theories has some aspects in line with inquiry-based learning approaches like SRPs and may support them. For instance, the open approach and the problem-solving approach, both well known and widely shared in Japan, emphasise autonomous activities by students, as a response to classical teaching approaches where students are passive recipients of instruction. The "mathematics learning incorporating 'question' as the main axis" approach, however, argues that the problem-solving approach is insufficient and instead emphasises the importance of taking up questions raised by students during mathematical activities and tackling them in a whole-classroom setting. The approach of researcher-like activity proposes the implementation in teaching and learning of different activities characteristic of actual scientists, such as reviewing papers, giving presentations, mounting panel discussions, etc. While the question of inquiry is not dealt with as a crucial element and the way of characterising scientists' work is naïve, Ichikawa's approach shares considerable common ground with SRPs. All of these approaches are published in books widely accessible to mathematics teachers in bookstores. Such resources, facilitating different instructional theories or approaches, would help to accommodate approaches like inquiry-based learning or SRPs.

4.6 Didactic infrastructure in Japan

The didactic infrastructure is the system of conditions supporting teachers' didactic activities in the classroom and consists of various different elements. One of them is the technological equipment in the classroom, including graphing calculators as well as computers with (for instance) spreadsheets, dynamic geometry software or CAS. The internet is a crucial element of such infrastructure, and the Japanese government has been promoting the use of ICT in the classroom; today, ICT elements such as computers, tablets, video-projectors, electronic blackboards and so on are more or less standard features in schools.[5] The use of tablets, in particular, allows easy access to the internet in regular classrooms, without having to move to a computer room. However, the actual use of these technologies in mathematics classes is very limited: graphing calculators, dynamic geometry software and spreadsheets, which are often used in schools in other countries, are rarely used in Japan, and Japanese mathematics textbooks do not suppose their use despite long-standing efforts to foster it.[6] Other constraints hinder this sort of inquiry in mathematics classes, even though the didactic infrastructure in Japan could support it. This situation also suggests that there are a few explorative activities in upper secondary school mathematics class.

Another element of infrastructure more invisible than the classroom environment is the broader educational system, which conditions teaching and learning in the classroom. We have already mentioned the subject Periods of Multidisciplinary Studies, the course Application of Mathematics and the task-oriented learning approach as vehicles for realising mathematical activities; these are situated at the three different levels of the Japanese educational system (subject, course and domain). This system, as part of the didactic infrastructure, contains within it possible places for organising SRP or inquiry-based learning. In the Japanese educational system one may also identify another, a higher-level element with conditions for SRPs: the Super Science High Schools (SSH) system, which aims to develop skilled international workers in the domain of science and technology (Super Science High Schools, n.d.). MEXT designates some upper secondary schools as sites to support the development of original curricula promoting scientific activities focusing on the sciences and mathematics. That is to say, outside of all subjects and the national curriculum, schools can organise special classes devoted to inquiry. While it is limited to particular schools with relatively advanced students, this system supports the implementation of inquiry of SRPs.

5 Comparison of conditions and constraints: Denmark and Kapan

In this section we compare the conditions and constraints that affect the realisation of SRPs in Denmark and Japan and reflect on the results of the analysis, which occurs around the six variables used above.

5.1 The official regulation of mathematics education

In both Denmark and Japan, national curricula officially regulate the way mathematics teaching is to be realised in the classroom. These curricula are written, in both countries, by a group of people from different professions: politicians, ministry officials, mathematics teachers, mathematicians, mathematics educators, and so on. This means that the finalised curricula are the result of negotiation among different parties with different perspectives on mathematics education. For example, politicians are often concerned with the nation's social and economic situation and how mathematics education can contribute to it; mathematicians, with the consistency of a body of mathematical knowledge to be taught step by step; etc. Among other factors, the results of international surveys like PISA and international movements in education, which today are often oriented towards the role of mathematics in real life, are elements many of these group members are concerned with. In

Denmark, this focus gives rise to the inclusion of new topics, probability theory and discrete mathematics; in Japan, to the creation of a new course, Application of Mathematics. Moves of this kind should support the realisation of SRPs. By contrast, in both countries the existence of "monuments" of predetermined knowledge to be imparted is a constraint on the implementation of SRPs. Emphasising the uses of mathematics in real life, the national curricula tend to include mathematical knowledge that will be useful in students' later lives. As these facts suggest, both countries' national regulations are very complicated and affect the situation in complex ways that further research may reveal.

5.2 Inquiry-based mathematics education and curriculum

The emphasis on "mathematical modelling" in Denmark and "mathematical activity" in Japan reflects an interest in inquiry-based mathematics education that may foster the implementation of SRPs in both countries. These keywords express the idea that mathematical knowledge and competencies must be acquired through the application of mathematics to some concrete situation. However, they also convey different perspectives on the notion of inquiry-based mathematics education: the Danish curriculum emphasises the application of mathematics to extra-mathematical situations, while the Japanese curriculum emphasises self-directed activities within mathematics per se (except in Application of Mathematics). From this point of view, Denmark may better support SRP than Japan.

However, as we mentioned above, the application of mathematics connects deeply with the perspective of *applicationism*, which becomes a constraint on the large-scale diffusion of mathematical modelling activities because it considers mathematical situations and extra-mathematical situations as completely separate entities (Barquero *et al.*, 2013). Teachers working from this epistemological viewpoint usually believe that any mathematical model should be first simply (more or less) mastered in the abstract, without any problematic questions being asked, and then just "applied" straightforwardly to extra-mathematical or other mathematical systems. Thus, the inquiry-based dimension of Danish and Japanese curricula both fosters and constrains SRPs.

In the Danish curriculum, a "study line project" is set up for multidisciplinary inquiries involving the application of mathematics. In the Japanese curriculum, very similarly, there is the Period for Multidisciplinary Studies course, which is supposed to involve cross-disciplinary activities. In both countries, these project-based studies are appropriate places for the implementation of SRPs.

5.3 Textbooks

Upper secondary school textbooks in Denmark and Japan are organised on the same lines: first the introduction of new mathematical concepts and then their application in exercise problems. This approach arguably shapes the teaching process in the classroom, in particular in Japan, where the use of designated textbooks is an obligation under the law; it may also reinforce applicationism and prevent the implementation of SRPs, as discussed above. However, textbooks also play a critical role as informational media that can enrich inquiry learning if used in its context, as it is done in Denmark; that is to say, the textbook is a condition or element of the didactic infrastructure that helps realise a rich study process.

5.4 Examinations and assessment

A constraint on SRPs common to both countries is the convention of high-stake examinations. In principle, SRP is concerned with students' own questions, which are not often investigated in exams. When students prepare for high-stake examinations, they are usually encouraged to ask instrumental questions: how do I get a high score? What kinds of problems were posed in past tests? What kinds of answer are likely to get higher marks? Under such circumstances, students' study and research processes are likely to be limited to those that can help them solve pre-set problems in the paper-and-pencil environment using techniques mastered beforehand. This is especially the case in Japan. The Danish examination system is actually in the process of reform, adopting a system that could possibly promote inquiry via elements such as report-writing, use of ICT, oral examination, group work, investigation of students' questions, and so on. In Japan, by contrast, the use of ICT in secondary education is very restricted, and oral exams and group work are not used as part of entrance exams. The examination is thus essentially a constraint on SRP-based teaching in Japan but is becoming a condition for supporting it in Denmark.

5.5 Teacher education system and resource system

Teachers' didactic equipment for teaching mathematics is optimised to a teaching approach based on monumentalism. In Denmark and Japan, the training for upper secondary school teachers presupposes explicitly or implicitly that they will deliver didactic knowledge under the paradigm of visiting works. For the future it will be necessary to explicitly look at the role of inquiry-based teaching in pre-service and in-service teacher training and see whether these create favourable conditions for implementing SRP at school level. The teacher education system in Demark, where a master's degree is a requirement for high school teachers, could have the effect of supporting inquiry-based teaching, since engaging in inquiry as part of a master's thesis in mathematics or mathematics education would help future teachers better understand inquiry processes and later implement them in the classroom.

5.6 Didactic infrastructure

The use of ICT as part of the didactic infrastructure is, in practical terms, a prerequisite for carrying out inquiry activities, including searching for established answers, data, research, etc., (thus) generating original answers by the inquirers and allowing them to derive new questions. The *dialectic of media and milieus* (see Glossary) without ICT greatly reduces the fruitfulness of study and research. In contrast with Denmark and probably many other countries (clashing with the widespread image of Japan as a country of high technology), Japanese mathematics education in upper secondary school has not yet met this criterion. Most mathematics teaching and learning still only occurs in a paper-and-pencil environment, and it is only outside mainline mathematics teaching, such as in the Period for Multidisciplinary Studies and Super Science High School curricula, that the use of ICT is promoted.

6 Final remarks

From a naïve perspective, one might expect very different conditions and constraints supporting or hindering the implementation of SRPs in upper secondary schools of each country, since there are many institutional differences at society and cultural level between Denmark and Japan. However, the analysis using our six variables shows more similarities than differences between the SRP ecologies in the two countries. In our study we mainly identified inquiry-based teaching approaches as a favourable condition for the realisation of SRPs. There were also some common constraints: entrance examinations and teachers' scholarship, which are optimised for the paradigm of visiting works. In both cases, introducing the application of mathematics to curricula provides a niche for inquiry activities. At the same time, emphasis on the application of mathematics tends to induce applicationism in the teaching, which we consider a constraint for SRP-based teaching.

As we have illustrated in this chapter, the conditions and constraints of SRPs originate from different institutions and constitute a very complex system to navigate. In neither Denmark nor Japan have SRPs so far flourished in upper secondary school classrooms. Some teaching experiments in SRP have been implemented in both countries, but SRPs in non-experimental situations probably do not exist. To study further whether other conditions could favour or preclude the large-scale realisation of SRPs in our societies as well as others is a crucial question for the ATD community around the world.

Notes

1 The two countries have rather different educational systems, which again differ from the French system where the design tool emerged. Therefore the countries are considered suitable for this analysis.
2 A Computer Algebra System is a computer programme able to perform algebraic manipulations and part of what is called ICT.
3 European Credit Transfer and accumulation System introduced by the EU as part of the ERASMUS programme. Sixty ECTS points corresponds to one year of full-time university studies.
4 The next version was published in 2018, but we use the term *current* for the 2009 version in this chapter. This is because the 2018 version was not implemented at the time of writing.

5   See for example the results of a national survey on ICT equipment in schools by MEXT: www.mext.go.jp/a_menu/shotou/zyouhou/1287351.htm.
6   See for example the website of T^3 Japan (Teachers Teaching with Technology): www. t3japan.gr.jp/.

References

Artigue, M., & Blomhøj, M. (2013). Conceptualizing inquiry-based education in mathematics. *ZDM Mathematics Education*, 45(6), 797–810.

Asami-Johansson, Y. (2015). Designing mathematics lessons using Japanese problem-solving oriented lesson structure: A Swedish case study (Doctoral dissertation). Linköping University, Sweden.

Baba, T., Iwasaki, H., Ueda, A., & Date, F. (2012). Values in Japanese mathematics education: Their historical development. *ZDM Mathematics Education*, 44(1), 21–32.

Barquero, B., Bosch, M., & Gascón, J. (2013). The ecological dimension in the teaching of mathematical modeling at university. *Recherches en Didactique des Mathématiques*, 33(3), 307–338.

Becker, J. P., & Shimada, S. (eds) (1997). *The open-ended approach: A new proposal for teaching mathematics*. Virginia, VA: NCTM.

Blomhøj, M., & Kjeldsen, T. H. (2006). Teaching mathematical modelling through project work. *ZDM Mathematics Education*, 38(2), 163–177.

Brousseau, G. (2002). *Theory of didactical situations in mathematics*. Dordrecht, The Netherlands: Kluwer.

Chevallard, Y. (2002). Organiser l'étude 3. Écologie et régulations. In J.-L. Dorier, M. Artaud, M. Artigue, R. Berthelot & R. Floris (eds), *Actes de la 11e École d'Été de Didactique des Mathématiques* (pp. 41–56). Grenoble, France: La Pensée Sauvage.

Chevallard, Y. (2009). Remarques sur la notion d'infrastructure didactique et sur le rôle des PER. Lecture given at the Journées Ampère in Lyon, France, May 2009. Retrieved from: http://yves.chevallard.free.fr/spip/spip/IMG/pdf/Infrastructure_didactique_PER.pdf.

Chevallard, Y. (2015). Teaching mathematics in tomorrow's society: A case for an oncoming counter paradigm. In S. J. Cho (ed.), *Proceedings of the 12th International Congress on Mathematical Education* (pp. 173–188). Cham, Switzerland: Springer.

Clausen, F., Schomacker, G., & Tolnø, J. (2010). *Gyldendals Gymnasiematematik, Grundbog B1*. Copenhagen: Gyldendal Uddannelse.

Danish Ministry of Higher Education and Science (2006). Retningslinjer for universitetsuddannelser rettet mod undervisning i de gymnasiale uddannelser. Retrieved from: www.retsinformation.dk/Forms/R0710.aspx?id=29265.

Danish Ministry of Education (2014). Bekendtgørelse om Pædagogikum I de Gymnasiale uddannelser. Retrieved from: www.retsinformation.dk/forms/r0710.aspx?id=163045.

Danish Ministry of Education (2016a). Aftale mellem regeringen, Socialdemokraterne, Dansk Folkeparti, Liberal Alliance, Det Radikale Venstre, Socialistisk Folkeparti og Det Konservative Folkeparti om styrkede gymnasiale uddannelser. Retrieved from: www.em u.dk/sites/default/files/gymnasiereform_1.pdf.

Danish Ministry of Education (2016b). Bekendtgørelse om Prøver og Eksamen i de Almene og Studieforberedende ungdoms- og voksenuddannelser. Retrieved from: www.retsin formation.dk/forms/r0710.aspx?id=179722.

Danish Ministry of Education (2017a). Bekendtgørelse om de gymnasiale uddannelser, Bek. Nr. 497. Retrieved from: www.retsinformation.dk/Forms/R0710.aspx?id=191190.

Danish Ministry of Education (2017b). Matematik B – stx. Bilag 122. Retrieved from: http://uvm.dk/-/media/filer/uvm/gym-laereplaner-2017/stx/matematik-b-stx-august-2017.pdf?la=da.

Dorier, J-.L., & García, F. J. (2013). Challenges and opportunities for the implementation of inquiry-based learning in day-to-day teaching. *ZDM Mathematics Education*, 45(6), 837–849.

Grøn, B., Felsager, B., Bruun, B., & Lyndrup, O. (2012). *Hvad er matematik? B.* [What is mathematics? B] Copenhagen: L&R Uddannelse.

Grønbæk, N., Rasmussen, A.-B., Skott, C. K., Bang-Jensen, J., Jensen, K. B. S., Fajstrup, L. ... Markvorsen, S. (2017). *Matematikkommissionen – afrapportering* [Report by the Mathematics Commission]. Copenhagen: Ministry of Education.

Ichikawa, S. (1998). *Hirakareta manabi heno syuppatsu: 21seiki no gakkō no yakuwari.* [Towards opened learning: The role of school in the 21st century]. Tokyo: Kaneko shobō.

Jessen, B. E. (2014). How can study and research paths contribute to the teaching of mathematics in an interdisciplinary settings? *Annales de Didactique et des Sciences Cognitives*, 19, 199–224.

Jessen, B. E. (2016). The collective aspect of implementing study and research paths–the Danish case. In M. Achiam & C. Winsløw (eds), *Educational design in math and science: The collective aspect* (pp. 40–46). Copenhagen: Department of Science Education, University of Copenhagen.

Jessen, B. E. (2017a). How to generate autonomous questioning in secondary mathematics teaching? *Recherches en Didactique des Mathématiques*, 37(2/3),217–245.

Jessen, B. E. (2017b). Study and Research Paths at Upper Secondary Education – a Praxeological and Explorative Study (Doctoral dissertation). University of Copenhagen, Denmark.

Jessen, B. E., Holm, C., & Winsløw, C. (2015). Matematikudredningen: Udredning af den gymnasiale matematiks rolle og udviklingsbehov [The Mathematics Evaluation: Role and need for development of upper secondary mathematics]. *IND's Skriftserie*, 42.

MEXT. (2009a). *Upper secondary school course of study* [in Japanese]. Retrieved from: www.mext.go.jp/a_menu/shotou/new-cs/youryou/kou/kou.pdf.

MEXT. (2009b). *Guideline for upper secondary school course of study: Mathematics* [in Japanese] Retrieved from: www.mext.go.jp/component/a_menu/education/micro_detail/__icsFiles/afieldfile/2012/06/06/1282000_5.pdf.

MEXT. (2009c). *Guideline for upper secondary school course of study: Periods of multidisciplinary studies* [in Japanese] Retrieved from: www.mext.go.jp/component/a_menu/education/micro_detail/__icsFiles/afieldfile/2010/01/29/1282000_19.pdf.

MEXT. (2013). *Periods of multidisciplinary studies to develop the ability required today (upper secondary school).* Tokyo: Kyōiku-shuppan.

Miyakawa, T., & Winsløw, C. (2009). Didactical designs for students' proportional reasoning: An "open approach" lesson and a "fundamental situation". *Educational Studies in Mathematics*, 72(2), 199–218.

Miyakawa, T., & Winsløw, C. (2013). Developing mathematics teacher knowledge: The paradidactic infrastructure of "open lesson" in Japan. *Journal of Mathematics Teacher Education*, 16(3), 185–209.

Nohda, N. (1991). Paradigm of the "open-approach" method in mathematics teaching: Focus on mathematical problem solving. *Zentralblatt für Didaktik der Mathematik*, 23(2), 32–37.

Ohara, Y. (2007). How are curriculum standards improved and implemented? In M. Isoda, M. Stephens, Y. Ohara & T. Miyakawa (eds), *Japanese lesson study in mathematics: Its impact, diversity and potential for educational improvement* (pp. 30–35). Singapore: World Scientific Publishing.

Okamoto, K., & Morozumi, T. (2008). *"Toi" wo jiku tosita sansū gakusyū*. [Elementary school mathematics learning incorporating "question" as a main axis]. Tokyo: Kyōiku syuppan.

Otaki, K., Miyakawa, T., & Hamanaka, H. (2016). Proving activities in inquiries using the internet. In C. Csíkos, A. Rausch & J. Szitányi (eds), *Proceedings of the 40th Conference of the International Group for the Psychology of Mathematics Education* (pp. 11–18). Szeged, Hungary: University of Szeged.

Schoenfeld, A. H. (1985). *Mathematical problem solving*. San Diego, CA: Academic Press.

Schoenfeld, A. H. (1988). When good teaching leads to bad results: The disasters of 'well-taught' mathematics courses. *Educational Psychologist*, 23(2), 145–166.

Skovsmose, O., & Nielsen, L. (1996). Critical mathematics education. In A. J. Bishop, K. Clement, C. Keitel, J. Kilpatrick & C. Laborde (eds), *International handbook of mathematics education* (pp. 1257–1288). Dordrecht, Netherlands: Springer.

Souma, K. (1997). *Sūgakuka Mondaikaiketsu no Jugyō* [The problem solving approach—the subject of mathematics]. Tokyo: Meijitosho.

Super Science High Schools (n.d.). In *Japan Science and Technology Agency*. Retrieved from: https://ssh.jst.go.jp (in Japanese).

Takahashi, Y. *et al.* (2015). *Shōsetsu Sūgaku I* [Detailed Mathematics I]. Osaka, Japan: Keirinkan. (in Japanese).

Winsløw, C. (2011). A comparative perspective on teacher collaboration: The cases of lesson study in Japan and of multidisciplinary teaching in Denmark. In G. Gueudet, B. Pepin & L. Trouche (eds), *From text to 'lived' resources: Mathematics curriculum materials and teacher development* (pp. 291–304). Dordrecht, Netherlands: Springer.

Winsløw, C., Matheron, Y., & Mercier, A. (2013). Study and research courses as an epistemological model for didactics. *Educational Studies in Mathematics*, 83(2), 267–284.

9

THE NEED FOR NEW TEACHING PRAXEOLOGIES IN THE PARADIGM OF QUESTIONING THE WORLD

Jean-Pierre Bourgade, Karine Bernad and Yves Matheron

1 Introduction

We study issues related to the dissemination and reception of study and research paths (SRPs, see Glossary) (Chevallard, 2007). The notion of SRP, as an extension of the concept of study and research activity (see Glossary), has emerged within the ATD in order to contribute to the development of teaching formats that make it possible to establish a functional relationship of individuals to knowledge. Indeed, in France, the paradigm of visiting works (Chevallard, 2002) is dominant in those institutions where a "teacher pedagogy" (Marietti 2009; Chevallard 2007) is widespread: the raison d'être of the knowledge to be studied is most likely not encountered by students; the students' *topos* (see Glossary), or even their role in the study process, is limited. The implementation of SRAs includes the formulation of so-called generating questions, the study of which leads to the encounter of a certain number of mathematical organisations that function as tools for producing answers to the initial question, as well as to other sub-questions arising from that place. Setting such study processes in "ordinary" classes raises a number of issues (Matheron & Noirfalise, 2011), essentially due to the discrepancy between these processes and classical didactic forms, which are typical of the "pedagogy of teachers" (like lecture-based courses, interactive lectures, teaching through imitation, etc.). Typically, high school students in France are expected to neither elaborate new techniques nor justify their efficacy: this falls into the teacher's *topos*.

With regard to the emergence of questions, students rarely take responsibility for studying them independently of the teacher's didactic intentions. Besides, within the framework of the paradigm of questioning the world, a "pedagogy of inquiry" would give a critical role to questions and the *topos* of students. Indeed, in this case, the aim is to collectively elaborate answers to a question, without prejudging which works might be crossed in the course of its study. As Marietti (2009) puts it:

Today, the transition from the paradigm of visiting works and from a teacher pedagogy to a "questioning" paradigm served by an adequate "pedagogy of inquiry" constitutes an open problem, and an important challenge is not only to train teachers in the pedagogy of inquiry, but also to identify praxeological needs for the conception and management of SRPs, and to disseminate praxeologies that fit these needs in the profession of teacher.

(p. 85, our translation)

The difference between paradigms can be made sharper: an SRP is generated by a genuine question Q and, in some sense, it is finalised by the question *itself*. Therefore, if the answer is (or includes) a given praxeology, this generating question Q provides a strong purpose, a *raison d'être*, to this praxeology. This, and the difficulty in developing SRPs in some institutions led to the notion of a *praxeologically finalised SRP* (or, for short, *finalised SRP*): while the generating question of the process remains prominent, it is generally chosen in order to facilitate or necessitate the encounter with a priori chosen works—mathematical works, for instance. In the paradigm of visiting works, the answer to the question Q is built by the teacher before students even enter in the study of Q (if they ever meet Q).

In contrast, in the case of the paradigm of questioning the world, the *milieu* of a genuine inquiry will progressively be enriched by questions Q_i derived from the initial question, available answers $A_j^\diamond$ (found on the internet, in books, provided by teachers or students, etc.), a variety of works W_k that may be useful for the study of the initial question. In particular, the teacher, indeed rather a *study helper*, does not impose the study of a given answer $A^\diamond$, or, if she does, this answer as any other must be submitted to the scrutiny of students: produced by a *media* (the teacher) among others, this answer has to be confronted to various *milieus* (*e.g.* calculators, books, other students, etc.) in a *media-milieu dialectic*. Such SRPs are called *praxeologically open SRPs*, or, for short, *open SRPs*.

Furthermore, the model of *didactic moments* (Chevallard, 2002; see Glossary) can also be used to study any process of construction or dissemination of professional praxeologies related to the implementation of SRPs. In a "pedagogy of teachers", students are seldom given the responsibility to explore the types of tasks and produce the techniques they have to study, let alone justify them. On the contrary, in a "pedagogy of inquiry", students are assigned part of the responsibility in the study of a question; they are also given a broader *topos* in the lead of the study. Specifically, the production of a rich enough *milieu* facilitates the emergence of a *media-milieu dialectics*: students have to assess the reliability of every answer raised in the process of study by checking it against the available milieu. In turn, such answers (accepted or rejected) enrich the milieu, etc. Since they have more autonomy in these dialectic processes, the students are more implicated in the production of techniques (exploratory moment) and the justifications of these techniques (technological-theoretical moment).

In this chapter we present two didactic analyses of the exploration of different professional types of tasks related to the implementation of finalised and open

SRPs, paying special attention to the exchanges between teachers and researchers working in the same team. In the next section we present the case of a team of researchers and a teacher implementing a finalised SRP at secondary level about the introduction of negative numbers. We study *conditions and constraints* on the internal didactic transposition's process undertaken by the teacher, starting from the appropriation of a provided document describing an SRP scenario, up to its implementation in the classroom. In the third section, we study the dissemination of a technique for the conception and phrasing of generating questions for open SRPs. We observe how university teachers with no previous experience in the "pedagogy of inquiry" take hold of this praxeological equipment. Finally, from the two cases we extract some conclusions about the dissemination of the pedagogy of inquiry within the teaching profession.

2 Dissemination of a finalised SRP in high school

In 2005, a national investigation programme, Activités Mathématiques et Parcours d'Étude et de Recherche dans l'Enseignement Secondaire (AMPERES), was initiated by the Commission Inter IREM[1] Didactique and supported by the French Institute of Education (IFÉ-ENS of Lyon). The aim was to make the mathematics of the programme live as by-products of answers to questions, which would provide them with rationales, "raisons d'être". Praxeologically finalised SRPs are considered a means to facilitate an evolution of ordinary teaching praxeologies. Teaching proposals are elaborated in an iterative process, alternating phases of implementation and phases of didactic analysis. In this inquiry we focus on issues related to the ecology of finalised SRPs—that is, on the study of the conditions and constraints that facilitate or inhibit the dissemination of SRPs in ordinary classes, in which mathematics from official programmes are taught and no systematic observation of practices is developed. This investigation develops a clinical study (Bernad, 2017) to analyse the didactic techniques activated by a teacher in order to carry out the exploratory and technologico-theoretical didactic moments—that is, the exploration of the type of tasks, the emergence of a technique to realise it, and the production of justifications of this technique.

2.1 An observation device

Since 2012, three teachers in the same high school in Marseilles, involved in the AMPERES project with Yves Matheron and Karine Bernad, were part of the secondary school Associated Educational Place (LeA, in French) Collège Marseilleveyre. Every other week, control meetings were organised in the institution. The three teachers and two didacticians worked together to study the following question: is it possible to use SRPs to teach all of, or part of, the curriculum of a given class? According to Yves Matheron & Serge Quilio (2015, p. 84), such cooperation must be anchored by what "systemic conditions and constraints allow, on what is expected to be possible for teachers, on what they have built based on their professional experience, and on what they could do with it". Thus, such a LeA generated *locally* new conditions under specific constraints for the teaching of mathematics: these new conditions hold for the setting up of SRPs, for instance.

One teacher, designated as y, joined this group in September 2014 to engage in the implementation of an SRP on the teaching of relative numbers in 5^e (12–13-year-old students). He was provided with a document, denoted: $W_{\text{SRP-LeA}}$. The letter W was chosen to represent the notion of work (Chevallard, 2003), defined in the ATD as "any intentional product of human activity". The fact that the LeA is a place both for experimentation and for the dissemination of this SRP explains the choice of the SRP-LeA index. This document describes the mathematical praxeologies (types of tasks, techniques and justifications) that must be elaborated by the students in the frame of the SRP, as well as some information about the techniques the teachers must use to manage the study and also the justifications of these techniques (technological-theoretical elements). The generating question, on which the SRP is based, is provided and the reasons for its choice are explained. It is important to emphasise that y did not contribute to the elaboration of this SRP.

Our analysis is based on the observation of class sessions and control sessions (both videotaped), individual interviews with the teachers (before and after the implementation of the SRP) and two documents that y produced to build an answer to the (possible) questions raised by the implementation of the SRP. In this investigation, the information available in $W_{\text{SRP-LeA}}$ is contrasted to these resources with the following question in mind:

What does $W_{\text{SRP-LeA}}$ provide the teacher with to help him implement the mathematics to be taught? What is made explicit? What adaptations of $W_{\text{SRP-LeA}}$ does he operate to produce his own organisation of the study?

2.2 Outline of an SRP about relative numbers

The SRP under consideration is grounded on the idea that relative numbers can be introduced as shortcuts for calculation programmes (see Chapter 5). This idea is based on the construction of a reference praxeological model (Gascón, 2014) as a link that ensures an epistemological coherence in the transpositive chain between mathematics and mathematics to teach. One of the "scholarly" mathematical constructions of $\mathbb{Z}$ can be briefly described as follows. It is necessary to have previously defined addition in $\mathbb{Z}$ as well as to have established its properties: associativity, commutativity, and regularity of the integers. Then we consider an equivalence relation R in $N \times N$, defined by $(n, m) \, R \, (n', m') \Leftrightarrow n + m' = n' + m$. Then we proceed to define $\mathbb{Z}$ as the set of equivalence classes for R. In our reference model, we define a function from $\mathbb{Z}$ to $\mathbb{Z}$, called "operator" and noted $O_{(b,c)}$ with $(b, c) \in N \times N$, as follows:

— if $b \geq c$ then $O_{(b, c)}(x) = x + (b - c)$; we say that $O_{(b, c)}$ is an additive operator,
— if $b < c$ then $O_{(b, c)}(x) = x - (c - b)$; we say that $O_{(b, c)}$ is a subtractive operator.

In fact, an operator is a special kind of *calculation programme* (see chapter 5) adding to a positive number x the (positive or negative) difference $b - c$.

If we call Ω the set of the operators thus defined and Def. the domain of definition of operator O, the relation R' defined on $\Omega \times \Omega$ by:

$$O_{(a, \ b)} \ R' \ O_{(c \ d)} \Leftrightarrow \forall x \in \text{Def}.O_{(a, \ b)} \frown \text{Def}.O_{(c,d)}, \ O_{(a, \ b)} \ (x) = \ O_{(c,d)} \ (x)$$

is an equivalence relation. We thus construct, in the same way, relative integers as equivalence classes for R'.

For instance, number -1 (that is, $(0, 1)$) is a shortcut for, say, the following calculation programme: $x + 61 - 62$, which can be oralised as: "Take a number, add 61, subtract 62." The organisation of the study, as planned by the designers of the SRP, goes roughly as follows: students are asked to operate some calculations as fast as possible; one possible technique for most of the calculations is to consider them as calculation programmes and to find a simpler, equivalent calculation programme. For instance, $2650 + 219 - 215$ can be seen as a realisation of "add 219, then subtract 215"; an equivalent programme is "add 4", since "$+ 219 - 215 = + 4$" ("$+ 219$" reads "add 219"). At some point, this technique faces a difficulty since, when adding 61 then subtracting 62, one meets the following programme, "add 61, then subtract 62", which is equivalent to "add -1": negative numbers are required, which have not been constructed yet. Facing students with this difficulty is one of the goals of the SRP, to lead them to build a new technique—and, thereby, a new type of numbers. In particular, relative numbers are introduced as additive or subtractive operators, obtained as notably simpler representative elements of equivalence classes of calculation programmes: this is the rationale, the *raison d'être*, for relative numbers that is pushed forward in the SRP. The calculations under study are algebraically represented by the formula "$a + b - c$", and the students are faced only with situations where $a \geq c - b$ where a, b and c are nonnegative decimal numbers. An equivalent simpler (at least in some situations) calculation programme reads: "add $(b - c)$" (or: "first calculate $b - c$, then add it to the chosen number a").

2.3 Study of the exploration of a type of tasks and the emergence of a technique

The case we are developing in this section is focused on a specific part of the SRP outlined above. According to the document $W_{\text{SRP-LeA}}$, it is expected that the teacher will propose an individual working time to search four calculations such as:

[1st] $2650 + 219 - 215$; [2nd] $23 + 12.3 - 2.3$; [3rd] $4374 + 62$–61; [4th] $4374 + 61 - 62$.

The indications given in the document are the following:

Students will possibly keep on adding 1 in the [4th] calculation, missing the difference with previous calculations. They can be convinced of their mistake by comparing the [3rd] and [4th] calculations. If we add 62 to and then subtract 61 from the same number 4374, it is likely that we will not get the same result as if we had added one less and subtracted one more. Soon, other students will notice the impossibility of applying the previous technique.

Nevertheless, numbers in these calculations are the same as in the previous calculation, even though their location is different. This, observed by students or, if not, pointed out by the teacher, should lead to a comparison between the [3rd] and [4th] calculation programmes.

In the observed implementation we are reporting here, the teacher (y) gave the students around three minutes to work out these calculations individually. After that, y took full didactic responsibility for the comparison of the last two calculations:

> "[I]f I consider the last two calculations [$4374 + 62 - 61$ and $4374 + 61 - 62$], you said: 'you made a mistake, you put the same twice'. In fact, I didn't put the same. We realise that [...] if that one is equivalent to adding 1, intuitively, you feel like saying that that one is equivalent to removing 1, to subtracting 1."

The emerging technique is grounded on mere "intuition" and no analysis is made of the fact that one is led to subtract a greater number from a smaller one. The exploratory moment is soon stopped and mathematical activity is mainly dedicated to the development of technological aspects and to their institutionalisation. It comes out that, instead of letting students analyse the situation and compare the 3rd and 4th calculations, the teacher y not only performed this comparison but also produced the adapted technique ("if that one is equivalent to adding 1, intuitively, that one is equivalent to removing 1, to subtracting 1"). Next, as a consequence, a student tried to apply the new technique ("subtracting 1") to one of the first (non-problematic) calculations. Indeed, the moment of exploration of the type of tasks and the emergence of the technique was underdeveloped. The teacher indicated the problematic situation:

> "[I]f you want, to summarise, when the number is greater, when we add more than we subtract, it's easy; it's what we've been doing since last week. That's what you needed!"

The student had not identified the problematic aspect of the task in such situations where $c > b$: the new technical element appeared to him as a completely new technique that could be applied in any situation, independently of the relationship between b and c. Ten minutes later, the teacher proceeded to institutionalise:

> "That was the problem, and that made our usual strategy fail."

Then he said:

> "I am only writing the steps of the calculation you just made in your mind. The goal is to write the outcome."

He repeated:

> "I am just describing the steps. The idea is to do a mental calculation."

The class activity is therefore set at the level of "mental calculation" when, in fact, the true goal of the SRP is to facilitate the emergence of relative numbers as representatives of classes of equivalent calculation programmes. This emergence requires the notion of "equivalent calculation programmes" and the notation "–1" as a shortcut for the calculation programme "subtracting 1". The teacher prevented this technological elaboration from happening by focusing on the students' giving their attention to mere "mental calculation". This is probably the consequence of the lack of didactic praxeologies necessary to manage the *media-milieu dialectic*. The observation of any crucial mathematical fact is not facilitated because the teacher provides the full technique and the elements of its justification. Therefore, students do not have the opportunity to compare different strategies or even to meet several specimens of the same type of tasks. In particular, this analysis reveals some difficulties in developing moments dedicated to the exploration of a problematic type of tasks and the emergence of a technique to realise it, and also moments devoted to the production of justifications for such a technique, thereby providing new definitions, theorems, etc. (that is, the technological-theoretical moment).

Another point is worth mentioning. In $W_{\text{SPR-LeA}}$, a technique is given to perform the mental calculation "$a + b - c$":

> One should oralise the calculation programme "$+ b - c$", calculate the difference d between b and c, if $b > c$, or between c and b, when $c > b$, then announce respectively: "add d" or "subtract d".

Using this technique is of paramount importance for the sake of the development of the SRP: it allows the students to meet calculation programmes and their equivalence. In particular, it facilitates the emergence of "– 1" as a notation for the calculation programme "subtract 1". However, in the observed sessions, the teacher indicated that this oralisation had no technical function for him. For instance, he said: "if we had to phrase our calculation, what would it be like?" Thus, he seemed to take phrasing as a justification rather than as a technical component. We are faced with what Marianna Bosch & Marie-Jeanne Perrin-Glorian (2013) evoke as effects of *logocentrism*, a term introduced by the French philosopher Jacques Derrida about the priority given to the oral discourse (*logos*) as a direct reflection of reasoning:

> Our culture tends to give speech a semiotic valence that seems to wipe out its instrumental valence, leading us to think of words as signs or signifiers of other objects, instead of also understanding them as technical entities, which are required to implement certain types of tasks—including the types of tasks that consist in pointing to new entities or producing new meanings.
>
> *(Bosch & Perrin-Glorian, 2013, p. 288, own translation)*

Once again, this should be contrasted with the indications provided in $W_{\text{SRP-LeA}}$:

> Then, it is the task of the teacher to indicate that a specific notation will be used to mean that, applying the result of the subsequent part of the calculation programme to the first number, is equivalent to subtracting 1. *The role of students* is restricted to the *justification of the notation* that comes as a simplification of the calculation programme.

In the next session, the teacher was supposed to bring a new notation along, such as "+ 45 − 46 = −1". He said to the students:

> "This is a notation I am showing to you, there is nothing to understand."

One minute later in the same session, y claimed:

> "Here [pointing to the sentence 'add 45 and subtract 46 is equivalent to subtracting 1' written on the blackboard], it is written in French, there [pointing to the equality '+ 45 − 46 = − 1'], it is written in maths."

According to y, this notation essentially sums up the sentence (it is therefore reduced to its *semiotic valence*—it only helps to refer to other objects). This can be seen as a pejoration of the *instrumental valence* of the written symbol "−1": the oral speech bears all meaning; the mathematical notation is just a *summary* of this speech. In fact, the notation "−1" bears the meaning of the calculation programme—which is its instrumental valence: adding n and then subtracting $n + 1$ is equivalent to adding −1; its practicality as a tool is more important than its mere semiotic valence. Besides, it is interesting to read the opposition between "in French" and "in maths", where "French" and "maths" are considered as competing *languages*. Logocentrism is arrived at when a process—the choice of a functional notation—is replaced by an arbitrary convention, like the arbitrariness of linguistic signs.

In any case, y seemed to neglect the possible justifications of the notation and, therefore, the meaning it can carry. As a consequence, he was not able to bring to life for his class a *media-milieu dialectic* that could lead students to propose notations, and at the end the notation is imposed on them as something arbitrary. The pejoration of oralisation as a technical element prevented students from analysing calculation programmes and their equivalence, thereby preventing them from having to denote calculation programmes by such notations as "−1". In an interview in June 2015, when asked to comment on the implementation of the SRP, y explained:

> I really felt this accordion-like understanding of students; there were times when I found some steps very easy, too easy: for instance, in the beginning, when working on calculation programmes. "I have never seen such an easy chapter," I heard that.

It is not clear how far the use of calculation programmes as a tool for the construction of the technological and theoretical elements (justification of the introduction of new numbers, of the choice of notation for negative numbers, of calculation rules on these numbers, etc.) was well understood: it seems to be considered as obvious, isolated from its possible use in the SRP. The teaching profession does not currently include in its praxeological equipment the appropriate praxeologies necessary for the implementation of adequate *media-milieu dialectics*. In particular, there is a lack of teaching strategies for helping students explore different types of tasks, construct techniques (use many specimens, let students compare their results and techniques, etc.) and propose technological-theoretical speeches (let student analyse different techniques, let them justify or invalidate techniques, etc.). As a consequence, the use of calculation programmes is perceived as a mere step in a process with loosely related phases.

These were the difficulties met in the cooperation process between researchers and a teacher in the implementation of a finalised SRP. We will now turn to another form of cooperation between a researcher and a team of university teachers to design an open SRP.

3 Initiating and managing a praxeologically open SRP

The first author of this chapter (hereinafter designated by the letter γ'), with the help of a didactician ξ, initiated the project to set up an *inquiry workshop* following the work developed at the Collège du Vieux-Port (Chevallard, 2011; Marietti, 2009), but at university level: the *prépa des INP* in Toulouse (France). In this institution, students of engineering sciences, aged 18, get a two-year intense training in science and humanities as a preparation for their access to engineering schools. Therein, pedagogy is mainly a "pedagogy of teachers", based on the will to "teach" rather than to help the students elaborate knowledge by themselves. γ' supported this workshop project for two years. After changing institutions, he was required to help a team of three teachers in the process of involving themselves in its pursuance. Two phases—building one's own praxeologies (design and management techniques, etc.) for the inquiry workshop, then disseminating them to colleagues—provided the opportunity to identify a series of difficulties encountered in the implementation of such a workshop. We present some of these difficulties and the effect they had on the remodelling of the teachers' professional praxeologies and, in particular, the emergence of a technique for the formulation of generating questions as well as some elements of a technology (justifications and groundings) for this technique.

The practical organisation of the workshop was as follows: 96 students were divided into groups of 24 students; each group was further subdivided into teams of four to five students. There were four 24-student groups, and two questions were asked (so two groups shared each question). The students were asked to elaborate an answer to the question, and they had 15 one-and-a-half-hour sessions (one every other week) to do it. Also, no a priori constraint was formulated regarding the tools used in the course of the work: internet, interviews with engineers or scholars, experimentations, software, books, questions to other teachers in the institution, etc.—"everything" was possible.

3.1 First encounter with, and identification of, a professional type of tasks

At first, γ' selected two questions. A high-school teacher proposed the first question after reading an article in a mathematics journal for teachers. The second one was elaborated by γ' after an inquiry based on the reading of a biographical article on Leonhard Euler. In the opinion of γ' at the time, both questions had the advantage of ensuring the encounter with some specific mathematical praxeologies (linear spaces, matrices, eigenvalues, etc.). More generally, all the ideas put forward by γ' were closely related to some mathematical praxeology. A few days before the beginning of the workshop, the two selected questions were:

> Q_1: Some photo editing software can sharpen blurry photos. How do they do it?
> Q_2: There are numerous constraints on the building of a bridge. In particular, the bridge is required to support heavy loads. How is it possible to foresee the maximum weight a bridge can withstand? (as formulated within the teacher's logbook, LB).

However, after a working session with the didactician ξ, the second question was abandoned in favour of the following, suggested by ξ:

> Q_2: Some mobile phones do not enter into standby mode until the user stops looking at them. How is this possible?

Obviously, ξ had something in mind when modifying the question: γ' understood it as a way of proposing a sharper question that would facilitate the start of the workshop (which may be one of the functions of a "generating" question). Also, this new question was not designed to ensure the study of mathematical themes—and even less, *chosen* mathematical themes.

In the following, we shall mainly focus on the moment of the first encounter with the following type of tasks:

> T_Q: Design a generating question for an inquiry workshop

—that is, on episodes where γ' is first confronted with T_Q and faces its realisation as problematic, and on the exploratory moment when γ' explores the type of tasks T_Q and seeks to elaborate a technique to fulfil it.

Surprisingly enough, the exploration of T_Q was *constantly* renewed over three years, not only during the periods devoted to the design of the generating questions. Though it was to be expected that T_Q would have paramount importance before the start of the workshop, it would be a serious mistake not to take into consideration the influence the question might have on the management of such workshops all the way through.

When y' started working with students on the exploration of Q_2, they were interested, even seduced: the freedom given to them to investigate in any direction and by any means, with great autonomy, was appealing. Nevertheless, a long series of incidents eventually reached its climax on the eve of the New Year break, when a team refused to work at all during the session, only to end with a provocative speech directed at y', mainly blaming y' for not providing help and refusing to be *precise regarding the sort of answers that were expected*. Other students had already complained: "What is it about? Do we have to programme a phone? Do we have to understand the engineers' programmes?" (LB); "we will never know whether the answer is satisfactory" (LB).

We can model this incident as follows. The workshop required that y' tackle the following type of tasks:

T_{MW}: Manage a group of students in the frame of an inquiry workshop

The kind of technique that y' elaborated required not giving a direct answer (yes or no) to the question "is our answer satisfactory?", rather asking questions about the elements of the submitted answer to allow students to identify weaknesses in their proposition (a way of initiating a media-milieu dialectics, where y' becomes an important part of the milieu). This technique raises an important issue: would it not be better if the students themselves were put in a position to ask these questions? This can be tracked in the logbook:

> y' handled the situation in compliance with the classical paradigm of study: he adopts the position of the bearer of knowledge [*that is, he uses a pedagogy of teachers…*] and establishes himself as the only judge of the adequacy of the submitted answer. By doing so, he can only comfort the team with the idea that he has definite expectations or even an idea of the answer, which he refuses to clarify to them. Another way to handle this little crisis would be to shift onto the students the responsibility of the construction of *the answer to the very question of the adequacy* of the answer […] y' could have suggested to this team that they put their ideas to the test by exposing them to other students from other teams.
>
> *(LB)*

Thus, y' was left with a significant difficulty regarding the management of the group: letting the team set their own "stopping criterion" leaves the students with a major *topos*—that is, more autonomy in managing the study since they decide by themselves whether the inquiry is over or not, but without the means to tackle the problem raised by teams that believe that they have a satisfactory answer (and that their criterion for satisfactoriness is satisfactory). Letting the students choose the stopping criterion can lead them to minimal criteria. All these questions are related to the difficulty of handling a satisfactory media-milieu dialectics (we see how γ' planned to include the other students into the milieu, etc.). This difficulty can be considered a problem not only for γ', but for the whole profession (Cirade, 2006; Chevallard & Cirade, 2010; Cirade, 2012) as it is progressively confronted by issues related to the new paradigm of questioning the world: implementing a genuine media-milieu dialectics does not belong to the professional types of tasks teachers are trained in. Here, it would have meant that the tentative answers elaborated by a team should have been considered as produced by a *media* (that is, by an instance which is not neutral regarding it) and therefore submitted to several *milieus* (that is, instances that behave as "pieces of nature" in this regard: without positive or negative intentions); for instance, another team could be asked to use the chosen team's answer to make something (write an algorithm, etc.): if failing to do so, this team could criticise the proposed answer which would have to be further elaborated, etc.

In his later exploration of T_Q, γ' took for granted that a successful realisation of T_{MW} relied on a satisfactory technique for performing T_Q. In other words, since the generating question itself can be considered as part of the milieu of its own study, the soundness of the media-milieu dialectics could depend on the inclusion of pieces of information in the generating question itself.

3.2 How to ask a question and what questions to ask

The exploration of T_Q by γ' raised the identification of at least two types of question: technical and technological. The difference relies on the use of distinct interrogative pronouns—or on the possibility of reformulating questions using one of the two pronouns, how and why. A "how" question is a technical question in which it is expected that, by answering it, the person describes a technique commonly used in a given institution to accomplish the task referred to in the question. A "why" question leads to an explanation (technology, in the sense of the ATD) of the technique alluded to in the question. Questions Q_1 and Q_2 were first analysed by γ' as being technological questions: though apparently "how" questions, they pulled the students towards the necessity of explaining why such or such a technique was used, or why such or such a device actually worked. To put it another way, γ' thought at first that the issues encountered in the management of the group originated in the fact that the questions asked for explanations (technologies) but that the students were provided with no a priori criteria for what kind of explanations would be admissible. Indeed, many teams proposed popularised explanations—leaving all technicalities unstated.

One could argue, though, that both questions are technical: Q_1 asks "how do they do it?", while Q_2 asks "how is it possible?". Answering the first question is to describe a technique used "to do it" in a given institution. To answer the second question, it is necessary to explain *why* a certain technique actually works ("how is it possible?" read as "why is it possible"); yet, to explain *why* something works, it is first necessary to show *how* it works. Therefore, the interpretation by γ' of his difficulties was not entirely satisfactory. Nevertheless, it is a milestone in the process of exploration of T_Q: the identification of the link between T_Q and T_{MW} indicates a certain direction for the elaboration of a technique for the realisation of T_Q, while the previous explanation (the questions were "technological"), though incorrect, is a technological element of the praxeology under construction.

3.3 Dissemination of a professional praxeology

After two years, γ' left *la prépa des INP* and a new team of teachers became responsible for the inquiry workshop: an English teacher (γ'_1), a mathematics teacher (γ'_2) and a physics teacher (γ'_3). None of them was acquainted with didactics of mathematics or with the pedagogy of inquiry—though the three of them had already done some important thinking on their professional (pedagogical) techniques. The four teachers met a couple of times to thoroughly discuss the groundings of the workshop: γ' warned them about the difficulties specific to the management of such a workshop; he also emphasised the possible relationship between these difficulties and the generating question of the SRP. They all agreed to work collectively on the production of questions after γ' gave some details of his own technique for T_Q. We will now briefly report on the process of formulating two questions for the workshop, as observed in the logbook of γ' and the emails exchanged with the other teachers.

First, as expected, γ'_2 had designed questions related to mathematics (or that would rapidly become mathematical problems). Consequently, γ' swept aside these questions by clarifying for γ'_2 the aims of the workshop (we find here the first elements of the technique elaborated by γ' in the first year of the workshop). After some days, two new questions arose:

> How to detect counterfeit artworks?
>   How to make artworks impossible to counterfeit?
>   Comments by γ'_1: "Problem: can the question asked to the students result in a catalogue of existing techniques […]? […] Up to what point should we investigate to make sure that the question provides a field of inquiry neither too wide nor too closed […], without investigating for them?"
>   *(Common logbook of $\gamma'_{1, 2, 3}$, 9/13/2015)*

The two questions are similar, and we see in the comments made by γ'_1 that part of the technology for T_Q has been acquired by γ'_i since they recognise the potential influence of the generating question on the ways the students might answer it. Here is a comment formulated by γ':

> I think we should find a wording that would allow the students to enter into an inquiry that would not finish rapidly in a catalogue of existing answers.

Ideally, we should find wording such as: "These days, it is estimated that about 50% of artworks on the market actually be counterfeit, despite the use of highly sophisticated scientific devices. Could you design and make a technical automated device that would ensure the authentication of an artwork?" The implication is to do "better" than existing solutions [...]. The writing "design and make" seeks to avoid the trap of an encyclopaedic catalogue of existing answers, and facilitates the answer to the question of the stopping criterion. [...] The question [...] should be converted to a question of the "could you do ..."-type.

(Email to y_i, 9/11/2015)

We see that y' not only proposes changes in the formulation of questions but gives rationales for these modifications: his comments reach a technological level, and he elaborates further his own technique by putting forward the following key tool: "convert questions to 'could you do ...' questions". This is an important modification of his own technique since questions Q_1 and Q_2 were not formulated this way. A few days later, y'_i wrote to y' about the questions they had produced:

Here they are:
Topic 1: To meet the energy needs of humanity, [how] can we use human beings themselves to produce daily useable energy?
Topic 2: 50% of works [currently circulating] in the art market might be counterfeit artworks:
How to detect counterfeit artworks? Is it possible to make artworks impossible to counterfeit? Give us your opinion, thanks.

(Email, $y'_{1, 2, 3}$ to y', 10/1/2015; bracketed words are modifications brought by y'_1 to a proposal of y'_2)

Though the two questions are already a distance from mathematics, and also match the additional demand that the questions should be *sharp* since both questions make clear reference to current sharp problems in our societies, they do *not* match the comment of y' regarding the importance of formulating "could you do ..." questions—even though we observe an attempt to "make technical" the questions by introducing the interrogative pronoun "how". The exploration of T_Q by y'_i shows a limitation in the elaboration of the technique expected by y':

As regards the two questions you have proposed, I feel they would benefit from the following rewritings:
1 To meet the energy needs of humanity, it can be contemplated to use the energy produced by human beings themselves. Could you suggest a device that would allow covering the needs in energy of the amphitheatre of *la prépa* using only (or mainly) the energy produced by its users?

2 Some sources claim that 50% of artworks circulating today on art market could be counterfeit: could you suggest a technique that would make impossible to counterfeit an artwork?

(Email from y' to y'ᵢ, 10/8/2015)

We see here that the new technique is used and we read "could you do ..." questions. Nevertheless, the technology of this technique is not well shared with y'_i:

Thanks for the questions, I feel they are indeed *more precise* with your modifications.

(Email from y'₁ to y', 10/8/2015, our emphasis)

Obviously, though the pedagogy of teachers focuses on questions, the evaluation of the quality of a question is generally based on clarity and precision requirements. In the pedagogy of inquiry, other questions can be asked about questions: about their capacity to generate inquiries, about the expectable range of these inquiries, etc. The *logos* of y includes the idea that the generating question of an SRP can play an important part in the development of media-milieu dialectic: the fact of using "could you do ..." questions is grounded on the hope that "doing" something will act as a *milieu*; whether the students can "do" something or not is like a "piece of nature" that will validate or not their answer according to their ability to use it to "do" something. This element can hardly be understood within the frame of pedagogy of teachers, where such a dialectics can barely exist. This operates as a constraint on the dissemination of praxeologies specific to the paradigm of questioning the world.

3.4 Emergence of the technique and contraction of the logos: more about generating questions

The previous section shows an evolution in the technique elaborated by y': not only does he insist on "could you do" questions, he makes more explicit the nature of "you" in "could you do": "you" refers not to an abstract person (as in philosophical dissertations) but to the subject of a given *institution*. In his seminar, Yves Chevallard (2010) had the same emphasis:

when [the institution] is elided [from the question] everything happens as if [the institution] were unique and as if there were a technique that is itself unique, and therefore implicitly universal, that answers the question. This is a language-effect that pushes back and hides the institutional relativity of praxeologies.

(Our translation)

Consequently, questions could be rewritten so as to mention the institution (we take the example of Q_1):

Q_{1bis}: What institutions dispose of a technique to sharpen blurry photos and what is this technique?

Following the technique proposed by y' to his colleagues, we could also propose a somewhat different question:

Q'_1: Some photo-editing software can sharpen blurry photos. Could you do it yourself on your computer?

This writing would translate as follows:

Q'_1: Some photo-editing software can sharpen blurry photos. In the institution of the inquiry workshop at *la prépa des INP*, is it possible to elaborate a technique to do it?

Now we can imagine a more general technique for the design of generating questions for inquiry workshops:

τ_Q: Ask a question that is deliberately not related to some specific field of knowledge or discipline studied in the institution where the workshop is to take place. The question must be designed independently of the desire to lead students to encounter a specific knowledge or theme. The question should be a sharp question. Finally, the question should be set at the level of a particular institution (for instance, the institution of the inquiry workshop itself).

The *logos* produced in the process of the dissemination of T_Q could be summed up as follows: the institution referred to in the question will be included in the milieu of the study. Consequently, the media–milieu dialectics will be easier to manage since the "stopping criterion" of the inquiry is grounded in the fact that any admissible answer will have to fit the needs of this institution (real or imaginary).

This *logos* explains something of the path followed by y' in the study of T_Q: his praxeological equipment is dominated (at least partially) by a teacher pedagogy that makes it difficult for him to handle media–milieu dialectics. The technique τ_Q "softens" the difficulty by anticipating it at the very level of the production of the generating question.

4 Conclusion

The prevailing pedagogy of teachers in the profession hampers the transition from a didactic paradigm focused on visiting works to a paradigm of questioning the world. The training of teachers to implement didactic devices such as open or finalised SRPs encounters this difficulty, even when didacticians lead the training. The lack of professional praxeologies for the generation and management of media–milieus dialectics appears to be an obstacle for the dissemination of pedagogy of inquiry. In this chapter, this was seen at two levels: both the performance of exploratory and technological-theoretical moments in a finalised SRP and the production of a generating question are problematic types of tasks for teachers, mainly because they are governed by a common *logos* that indicates the necessity for the teacher to *lead* the study. This is obvious in the situation explored in the second section of this work, but also in the one at the core of the third section since the missing *logos* of teachers about the functions of questioning sufficiently explains their inability to endorse a new technique for the production of generating questions.

Note

1 Institut de Recherche sur l'Enseignement des Mathématiques.

References

Bernad, K. (2017). Une contribution à l'étude de conditions et contraintes déterminant les pratiques enseignantes dans le cadre de mises en œuvre de parcours d'étude et de recherche en mathématiques au collège (Doctoral dissertation). Université d'Aix-Marseille, France.

Bosch, M., & Perrin-Glorian, M. J. (2013). Le langage dans les situations et les institutions : Essai de croisement de points de vue TAD et TS. In A. Bronner *et al.* (eds), *Questions vives en didactique des mathématiques : Problème de la profession d'enseignant, rôle du langage* (pp. 267–312). Grenoble, France: La Pensée Sauvage.

Chevallard, Y. (2002). Organiser l'étude: 1. Structures & fonctions. In J.-L. Dorier *et al.* (eds), *Actes de la 11e École d'Été de Didactique des Mathématiques* (pp. 3–32). Grenoble, France: La Pensée Sauvage.

Chevallard, Y. (2003). Approche anthropologique du rapport au savoir et didactique des mathématiques. In S. Maury & M. Caillot (eds), *Rapport au savoir et didactiques* (pp. 81–104). Paris: Éditions Fabert.

Chevallard, Y. (2007). Passé et présent de la théorie anthropologique du didactique. In L. Ruiz, A. Estepa & F. J. García (eds), *Sociedad, escuela y matemáticas: Aportaciones de la Teoría Antropológica de lo Didáctico* (705–746). Jaén, Spain: Universidad de Jaén.

Chevallard, Y. (2010). Journal du séminaire TAD/IDD. Retrieved from http://yves.chevallard.free.fr/spip/spip/IMG/pdf/journal-tad-idd-2009-2010-6.pdf.

Chevallard, Y. (2011). La notion d'ingénierie didactique, un concept à refonder: Questionnement et éléments de réponse à partir de la TAD. In C. Margolinas *et al.* (coord.), *En amont et en aval des ingénieries didactiques : XVe École d'Été de Didactique des Mathématiques* (pp. 81–108). Grenoble, France: La Pensée Sauvage.

Chevallard, Y., & Cirade, G. (2010). Les ressources manquantes comme problème professionnel. In G. Gueudet & L. Trouche (dir.), *Ressources vives: Le travail documentaire des professeurs en mathématiques* (pp. 41–55). Rennes, France: Presses universitaires de Rennes.

Cirade, G. (2006). Devenir professeur de mathématiques : entre problèmes de la profession et formation en IUFM : Les mathématiques comme problème professionnel (Doctoral dissertation). Université de Provence-Aix-Marseille I, France.

Cirade, G. (2012) La formation des professeurs: entre analyse de praxéologies professionnelles et étude de problèmes de la profession. In J. L. Dorier & S. Coutat (eds), *Enseignement des mathématiques et contrat social : Enjeux et défis pour le 21e siècle – Actes du colloque EMF2012* (GT2, pp. 314–323). Geneva, Switzerland: Université de Genève.

Gascón, J. (2014). Los modelos epistemológicos de referencia como instrumentos de emancipación de la didáctica y la historia de las matemáticas. *Educación Matemática, 25*(E), 99–123.

Marietti, J. (2009). Le concept de PER et sa réception actuelle en mathématiques et ailleurs: Une étude préparatoire (Mémoire de 1ère année de Master Sciences de l'Éducation). Université Aix-Marseille I, France.

Matheron, Y., & Noirfalise, R. (2011). Du développement vers la recherche: Quelques résultats, issus du projet (CD)AMPERES, relatifs à la mise en oeuvre de PER dans le système d'enseignement secondaire. In M. Bosch *et al.* (eds), *Un panorama de la TAD* (pp. 57–76). Barcelona, Spain: Centre de Recerca Matemàtica.

Matheron, Y., & Quilio, S. (2015). L'accès au milieu scolaire pour l'élaboration et l'expérimentation d'ingénieries didactiques de recherche: Conditions et contraintes. In A.-C. Mathé & E. Mounier (eds), *Actes du séminaire national de didactique des mathématiques 2014* (pp. 80–91). Bordeaux, France: ARDM & IREM Paris VII.

PART 4

ATD and the Teaching Profession

10

THE STUDY OF TEACHERS' MATHEMATICAL AND DIDACTIC PRAXEOLOGIES AS A TOOL FOR TEACHER EDUCATION

Gisèle Cirade and Anne Crumière

This short chapter has two sections: the first examines the notion of "teaching trade" and the difficulties that appear in the exercise of the trade; the second looks at the role of teacher training or teacher education in the evolution of this trade towards a more professional status.

1 The teaching trade and its difficulties

Before discussing the *training* of mathematics teachers, let us begin by talking about the *trade* of mathematics teaching based on Chevallard (2013). Having specified that the word *trade* would be used "in a sense specific to the ATD", the author states:

> The trade of mathematics teaching is what anyone in this trade must do. To practise the trade of general practitioner means, in particular, receiving patients in your office, questioning them, examining them, asking them for their health card, writing prescriptions; and it also means visiting patients at home, etc. Similarly, practising the trade of mathematics teaching leads to what I would call the gestures of the trade—for example, "teaching", "preparing your course", "giving homework", "correcting homework", "writing an answer sheet", "filling out school reports", etc.
>
> *(p. 86, our translation)*

This sketched list of *gestures of the trade* raises the question of the *praxeological equipment* of teachers. Obviously, the analysis of such equipment will depend strongly on the authority, personal or institutional, that proposes it. Let's take a quick example, in France, with the "referential of professional competences of teaching trades and education" (Ministère de l'Éducation Nationale, 2013), to situate what is happening from the employer's point of view. In the first part, we find the "competences common to all teachers and educational personnel"; there are 14 of them, we list five of them:

> 1. Share the values of the Republic; 3. Know the students and the learning processes; [...] 8. Use a foreign language in situations required by his trade; [...] 10. Cooperate as part of a team; [...] 14. Engage in an individual and collective approach to professional development.

This referential then proposes five "competencies common to all teachers", the last three being indicated as being addressed to them as "expert practitioners of learning":

> P1. Master disciplinary knowledge and its didactics
> P2. Master the French language as part of your teaching
> P3. Build, implement and facilitate teaching and learning situations that take into account the diversity of students
> P4. Organise and ensure a group operating mode that promotes student learning and socialisation
> P5. Assess student progress and achievements.
>
> *(Ibid., our translation)*

The introduction to the referential specifies that "each competence [...] is accompanied by items that detail its components and specify its scope"; for example, consider one of the five generic items accompanying competence P3:

> Know how to prepare the class sequences and, for this, define plans and progressions; identify objectives, content, devices, didactic obstacles, support strategies, training and assessing modalities.
>
> *(Ibid., our translation)*

Here we can identify an emblematic type of tasks: preparing a sequence, as well as another one, which is directly related to it: defining a plan and a progression. We also see the appearance of subtypes of tasks of the first one: identifying objectives, identifying contents, etc. Without pursuing the analysis of the above-mentioned professional competences, we note that we see the emergence of what will constitute, for the employer, the didactic stake in teacher training.

After introducing the concept of a *trade* in ATD, Yves Chevallard (2013, p. 86) specifies: "What matters now is this: the exercise of a trade encounters *difficulties*". It is these difficulties, and how to identify them, that we will now focus on. The question of identifying the difficulties encountered in the trade is crucial, and it is a delicate one. The dialectic of the individual and the collective plays an essential role here because it is through the difficulties experienced by *some* professionals that we will be able to identify those faced by *the* professionals—the former revealing the latter, which are in fact the main focus of our research.

Let us introduce a system aimed at bringing to light the difficulties encountered by teachers. Implemented in some initial teacher training courses under the name of the *questions of the week*, this system can be presented succinctly as follows:

> Every working week, during a working session with the entire cohort, student teachers, whether they are preparing for CAPES[1] in the first year or are trainee teachers in the second year, are invited to write down, individually, a difficulty they have encountered and the questions it raises for them.
>
> *(Cirade, 2006, p. 63, our translation)*

For example, we propose below some of the questions of the week formulated by a 2nd-year student teacher, Margot, throughout her teacher-training year 2014/15 (Cirade, 2006, pp. 119–133, the numbers correspond to the week of training). As can be seen, the difficulties encountered are diverse:

> 3. The majority of students in my class do their exercises but some do not. What to do about it? Punish them? // 7. I start a chapter on the configurations of the plan mainly based on reviewing parts of lower secondary mathematics to learn how to perform proofs. How to "schematise" a geometric proof? How can one explain a type of reasoning? Can we talk about logic? // 8. In terms of generalities about functions, what techniques should students know about function variation? My book only uses reading graphs. Is the "algebraic" method (showing that, if $a \leq b$ then $f(a) \leq f(b)$, etc.) still on the syllabus of grade 10? // 11. A student goes to the blackboard to solve a geometry exercise. He writes his solution, it is correct, but there is a faster way. One student points this out orally. Should this other reasoning be apparent on the blackboard? Do students have to take note of this? // 14. In the chapter Locating in the plan, it is recommended to make markings in a spreadsheet. How to create an activity about marking in a spreadsheet? What can be its mathematical interest? // 16. About the coordinates of a vector, the row data $\vec{u} = (a;\, b)$ and column data $\vec{u} = \binom{a}{b}$ appear in various documents. What is the "right" writing? // 22. Let ABC be any triangle. Note P the intersection point of the inner bisector of A with [BC]. How to prove that $PBPC = ABAC$? // 24. How to develop a "common" assignment at the end of the year?

We will not comment on these questions because we wish to emphasise the following: the difficulties mentioned by a particular *individual* (in this case Margot) are, in fact, an indication of the difficulties faced by *the collective* of student teachers and, beyond that, of the teaching *profession*. In the context of the *forum of questions and answers* where the *questions of the week* were approached, it was clearly mentioned to the students that the aim of the work was not answering *Margot*, but the *questions* she was brought from the field, which participants must learn to see as questions of the *profession*.

If the system of the *questions of the week* makes it possible to highlight the *difficulties* encountered in the trade, this is of course not the only way. The work carried out by André Pressiat (chapter 13) describes difficulties that have been identified in different contexts (visits of a teacher in training, meetings at the ministry, publications, etc.), even if they are usually ignored or at least not approached as difficulties as such. We now examine what can happen after identifying difficulties encountered by teachers, beginners or not, in the practice of their trade. To do this, let us look at how Chevallard (2013) models the inquiry that can occur from the moment a difficulty is recognised:

> A difficulty having been recognised by a person or institution ξ, it can be transformed, for a person or institution ξ^*, into a question to which an answer must be given. [...] The recognition that a difficulty affects the practice of the trade, its transmutation into a question Q, the construction of an answer A and the control of the validity and value of this answer are by definition within the noosphere of the trade.
>
> *(p. 88, our translation)*

Pressiat (2017) places itself de facto in the *noosphere* (see Glossary) of the trade when he identifies a difficulty and transforms it into a *question* to be studied. When the question is studied and disseminated, it acquires the status of a *problem for the noosphere of the trade*—that is, what is called, in short, a *problem of the profession*. The same applies, of course, to the system of the questions of the week in the context of initial teacher training. It should also be noted that, if we look at all the questions of the week asked over several years and focus on a particular difficulty, in some cases the questions of the week abound—for example those relating to homework (Cirade, 2011)—while in other cases they emerge sporadically—this is the case for those relating to alternating internal angles (Cirade, 2008).

Chevallard (2013) specifies what he calls the level of professionalisation of the noosphere of a trade or, in short, the *level of professionalisation of a trade*. We can start by assuming that the noosphere of a trade is a profession if it satisfies a certain number of criteria from which various lists have been drawn up in a convergent manner: full-time occupation, the establishment of training or university schools, a national association of professional ethics, state licensing laws, etc. Recalling the work carried out by the American sociologist Amitai Etzioni (1969), which makes it possible to define what is called *semi-professions* but also other levels of professionalisation of the trade, Chevallard mentions trades for which the noosphere does not satisfy these criteria:

> Let us add that, of course, there are trades, "small trades", whose degree of professionalism is lower, even much lower than that of semi-professions. Etzioni has thus created the term McJob (from McDonald's restaurants) to designate, according to the eponymous article in Wikipedia, jobs "where little training is required, staff turnover is high, and workers' activities are tightly regulated by managers".
>
> *(Chevallard, 2013, p. 90, our translation)*

To simplify, we use the term *profession problem* to refer to a problem dealt with in the noosphere of the trade, even if the *level of professionalization* of this noosphere is not such that it can be described as a profession. It should also be noted that "it is by assuming the difficulties of the teaching trade as *problems*, the resolution of which requires appropriate fundamental and applied research, that the noosphere of the teaching trade can give rise to a true profession" (ibid., p. 90, our translation).

2 The professionalisation of the trade

The question of building a "university vocational training" for teaching (Chevallard & Cirade, 2009) continues to be an insistent one, and it provides an opportunity to strongly reformulate the links between training and research:

> The university would fail in its mission if it only offered teacher training to *readers* in science that is safe, but without scope for contextual intelligibility and practical effectiveness *in professional matters*. We have to go backwards. Building appropriate answers $A^{\heartsuit}$ requires the development at new cost, without false economies, of teaching *techniques* that will be tested in a thousand ways, *technologies* that project on them an adequate intelligibility, rooted in *theories* that cannot be expected to exist ready-made in the realm of academic knowledge and that must, therefore, continue to be produced. The university must not only provide safe *lectores* for teacher training. It is vital that it accepts the exciting adventure of becoming a collective *auctor*, without arrogance, without boastfulness, with the generosity due to professions that, every day, contribute as much as possible to giving society its own intelligibility and each of its members the intelligence of the situations experienced.
>
> *(p. 55, our translation)*

We will now repeat the problem of the praxeological equipment of teachers, starting not from the competences of the referential mentioned in section 1, but from the questions to be studied in the didactic systems of teacher training that are mainly produced by the question Q_f: how to occupy the position of a teacher in the school didactic systems? (Artaud, Bourgade, Cirade & Sémidor, 2016). The authors specify that this implies identifying a certain number of types of tasks $T_{\pi i}$ to be performed in this position and answering the associated questions $Q_{\pi i}$: how to perform $T_{\pi i}$? The problem of the didactic transposition of the praxeologies to teach then arises, as in any didactic system, and is deployed in multiple sub-problems, among which we can mention: what are the questions to be studied? What are the answers to be provided? Where did they come from? What conditions and constraints influence the dissemination of questions and answers developed in institutions for the production of relevant knowledge, including research in didactics?

These are questions relating to *teaching* praxeologies, and similar questions can be stated for the *training* praxeologies: How to help the study community to study the questions $Q_{\pi i}$? This is the second problem trainers have to face, which is not unrelated to the previous one.

2.1 The difficulties of the trade of teacher educator

We will now consider a reflection on the *difficulties* encountered by *teacher trainers or educators* in the exercise of their trade. Until now we have essentially considered the *teaching* trade, but the developments proposed by Chevallard (2013) allow us to place ourselves within the general framework of the exercise of *a* trade. We choose here to place ourselves within the framework of the trade of *trainer* by now considering the reflection mentioned above:

> I have great difficulty in justifying the need to carry out a mathematical prax-eological analysis—to explain the types of tasks, techniques and technologies of a mathematical organisation—to students preparing for the teaching trade, as well as the didactic praxeological analysis and the relationship between the two. Students find this work excessive. How to conduct didactic analysis in the context of the ATD should be developed.
>
> *(Cirade, 2019, p. 6, our translation)*

In this statement, two aspects can be distinguished: first, the *raisons d'être* of a didactic analysis; second, the *technique* to be used to carry out a didactic analysis—it should obviously be noted that, beyond the technique to be used, it is the prax-eology itself that must be considered. In the immediate term, let us try to identify why it is useful, or even essential, to include didactic analysis in the praxeological equipment of future teachers. To study this question, we will place ourselves within the framework of the *primordial* problematic, which can be stated as follows:

> The *primordial* problematic is to determine, for a given instance (personal or institutional) u, the *praxeological equipment* which would be considered ade-quate, by an instance v (with possibly $v = u$), under constraints K and in con-ditions C, for the design and implementation of a given project Π. This type of questioning, therefore, aims to determine the set $\wp$ of praxeologies $\mathscr{P}$ [...] considered, by the authority v, useful or essential to u to engage in the project Π under the constraints K and in conditions C.
>
> *(Chevallard, 2017, pp. 31–32, our translation)*

Consider the case where $u = (I, p)$ is the institutional position of trainer in a teacher-training institution, the project Π consisting in training teachers and v a body composed of didacticians coming to occupy a certain position in the noo-sphere of the trainer's trade. The question is to determine the praxeologies $\wp$ useful or essential to carry out such a project—under constraints K and in conditions C which we will not detail here. In this didactic institution I, it is a question of setting up teaching praxeologies. A didactic system is created around the generating question Q_f: "How to occupy the position of teacher in school didactic systems?", with students X and study aids Y, to bring together a didactic milieu M and confront it in order to produce an answer $A^{\heartsuit}$. Other didactic systems will follow, around questions that will appear during the study and which, for some, may come from the transmutation of difficulties that will have been recognised in the training (see section 1.2). In any case, this study process can be modelled by the Herbartian schema (see Glossary) presented here in its semi-developed form:

$$[S(X;\ Y;\ Q) \curvearrowright M] \hookrightarrow A^{\heartsuit}$$

The didactic milieu, M, which is thus brought together is of course evolving, and it is throughout the study process that it is enriched. Let us take up again what Chevallard says (2019, pp. 22–23) about the creation of this milieu M which will make it possible to produce the answer $A^{\heartsuit}$:

> The first is the search—"in the literature" and, in particular, on the internet—for existing answers offered by other persons or institutions. Such answers are usually denoted by $A^{\lozenge}$ [...] At his stage, the milieu M is, therefore, to be written thus: $M = \{A_1{}^{\lozenge}, A_2{}^{\lozenge}, \ldots, A_m{}^{\lozenge}, \ldots\}$.
>
> To draw upon the answers $A_i{}^{\lozenge}$ ($1 \leq i \leq m$), the didactic system has recourse to works of various kinds, like theories, experiments, historiographical narratives, etc. Therefore the milieu is now to be written: $M = \{A_1{}^{\lozenge}, A_2{}^{\lozenge}, \ldots, A_m{}^{\lozenge}, W_{m+1}, W_{m+2}, \ldots, W_n, \ldots\}$. To use these works, the student [...] needs to study them. What does it mean to study a work W which is not itself a question? Such a study consists in studying a number of questions Q_w about the work under study [...] So that the milieu M takes on the following appearance: $M = \{A_1{}^{\lozenge}, A_2{}^{\lozenge}, \ldots, A_m{}^{\lozenge}, W_{m+1}, W_{m+2}, \ldots, W_n, Q_{n+1}, Q_{n+2}, \ldots, Q_p\}$.

For many of the didactic systems studied in training, an answer $A_i{}^{\lozenge}$ can be a class report, a textbook extract, a teacher's website, a document distributed to students, etc. These answers will be analysed and assessed to develop instructional products. In training, we will, therefore, have to analyse the teaching praxeologies, by approaching both the *praxeological analysis of the content to teach* and the *praxeological analysis of the teaching activities*, as mentioned in the reflection proposed at the beginning of section 2.1. More broadly, it is a question of *constituting and studying corpuses of works*, which can be of a diverse nature, and of integrating some *praxeologies* into the milieu M to study these *corpuses*.

2.2 Infrastructures for didactic analysis

We will now briefly present two examples based on our work carried out as part of teacher training processes at the ESPE Toulouse Midi-Pyrénées. The first example (Cirade, 2019, section 2.1) concerns the *mathematics to teach*, in the case of the theme entitled *Weighted averages* in the curriculum for the 4th grade (students aged 13–14) in force at the beginning of the 2009 school year in France. In training, a didactic system is built around a question Q such as "What is the mathematics to teach about 'weighted average'?" As the study progresses, the milieu M will be enriched with answers $A_i^\lozenge$—a curriculum extract, textbooks extracts, etc.—questions Q_k—"What does the syllabus and textbooks say about this?", "What is a weighted average?", etc.—as well as other works W_j. Among the works W_j there is a mathematics book on averages, but others are introduced that are less "visible" but essential because they are the ones that will allow the study of the above-mentioned documents (curriculum extract, textbooks extracts, etc.): the elements needed to perform a *praxeological analysis* of the mathematics to teach. This analysis will clarify what is problematic for student teachers—namely, what is a *weighted average* and why the programme introduces this notion. In this case, the questions addressed focused, in a non-independent way, on the four components of the praxeology under consideration. We could then see that the topic "Weighted averages" consists only of one type of tasks; that the differences between average and weighted average are of a technical nature and relative to the numerical series considered (their size and, above all, the way they are presented: as a list, a distribution table, a graph). We could also see that there were not two definitions, but a definition and a *property*, and that the property can be justified; that, in all cases, a series of values can be summarised by a number that equidistributes the values in the sample; etc. What matters here is that the initial question—"What is the mathematics to teach?"—can be studied *scientifically*: the notion of praxeology providing a *model* for analysing, and therefore better *understanding*, what is mentioned in the curriculum, what is found in textbooks, etc., and thus for changing the students' relations to the mathematical objects considered.

The second example (Crumière & Cirade, in press) is related to the block of content—a sector in the scale of codeterminacy (see Glossary)—called "Data organisation and management" in lower secondary school in France (students aged 12–15) at the beginning of the 2016 school year. Trainers from the ESPE Toulouse Midi-Pyrénées offered a workshop to the student teachers with the following instruction: "Choose a statement to start an 'activity' to teach a mathematical organisation around the type of tasks 'Determine a median'." Before this workshop, an analysis of the mathematical praxeologies of this sector was carried out with the teacher students. During the workshop, the students produced about 15 proposals—which were all answers $A_i^\lozenge$ to the question "How to determine a median?". The analysis of these responses revealed a number of difficulties. In fact, student teachers had the impression that, by offering highly guided activities, they achieved greater control over the time and progress of the study and, through this, better classroom management—which is often the main concern of beginning teachers. However, this management is very closely linked to the robustness of the proposed activities and, in particular, to the proper management of the progress of the study through crucial questions. The analysis of these responses $A_i^\lozenge$ also highlighted the absence of raisons d'être for the median in the work proposed by the student teachers. In addition, the analysis of the corresponding mathematical prax-eologies revealed the following points. Even if the sector "Data organisation and management" occupies a significant place in lower secondary school curricula in France, it is often ignored by the profession because it is considered to be of low mathematical level and therefore not a priority. Some technological elements are "non-stabilised", and there is persistent uncertainty regarding the basic elements of statistical models (data, values, variables, series, etc.), which confirms a certain pejoration of statistical knowledge. In conclusion, the didactic analysis revealed recurrent difficulties encountered by teachers in the exercise of their trade, even if the inquiry started from those encountered by a group of student teachers. Furthermore, it also provided some tools for the development of new teaching praxeologies.

2.3 Conditions and constraints

Within the framework of a training project aimed at encouraging the involvement of student teachers in the design of mathematical and didactic praxeologies in a *scientifically sound manner*, the question of *didactic infrastructures* in training arises, in the dual sense of the content at stake (teaching praxeologies) and of the direction of the study (training praxeologies). Many conditions and constraints must be taken into account, at the different levels of the scale of didactic codeterminacy (see Glossary). At the lowest levels of the scale, the conditions related to the themes, sectors, domains and the (mathematical) discipline itself should be further studied. However, upper-level conditions are also significant. For example, *the denial of the need for scientifically based praxeological equipment*, which seems to us to be essentially due to the *pejoration* of didactics in our societies. This constitutes in our opinion one of the most important constraints in the context of such a project because in the trade of teaching and, more generally, in the corresponding semi-profession, the situation described by Chevallard (1997) still seems to be relevant:

> we must note the absence of a language rich enough and widely enough shared to allow an objective (and not simply personal) analysis of even the most common professional situations, with consequently a weak collective and individual capa-city to communicate, debate, think even, about the objects of an activity that can easily be enclosed, for that, in the repetition of gestures and in technical solipsism.
>
> *(p. 23, our translation)*

This has a very significant impact on teacher training and creates many difficulties for trainers, such as the one mentioned in section 2.1, which points to the difficulty of justifying to students the need to rely on scientifically based tools—it should be noted that the question does not arise regarding the discipline they are supposed to teach.

A source of difficulties is obviously due to the complexity of the didactic stake in training—namely, *teaching praxeologies*. However, the low development of *teacher-training praxeologies* is of greater importance because it does not facilitate the creation and dissemination of solid teaching infrastructures for teacher training.

Note

1 In France, the Certificat d'Aptitude au Professorat de l'Enseignement du Second degré (CAPES) is the competition for the recruitment of certified teachers for general subjects. The winners of this competition then follow a second-year training course at the École Supérieure du Professorat et de l'Éducation (ESPE) as trainee teachers.

References

Artaud, M., Bourgade, J.-P., Cirade, G., & Sémidor, P. (2016). Analyser des praxéologies de formation: Apports de la TAD. Symposium conducted at the 4e colloque international de l'ARCD, "Analyses didactiques des pratiques d'enseignement et de formation: Quelles perspectives ?", Toulouse, France. https://arcd2016.sciencesconf.org/browse/session?session id=17180.

Chevallard, Y. (1997). Familière et problématique, la figure du professeur. *Recherches en Didactique des Mathématiques*, 17(3), 17–54.

Chevallard, Y. (2013). L'évolution du paradigme scolaire et le devenir des mathématiques : Questions vives et problèmes cruciaux. In A. Bronner *et al.* (eds), *Questions vives en didactique des mathématiques : Problèmes de la profession d'enseignant, rôle du langage* (pp. 85–120). Grenoble, France: La Pensée Sauvage.

Chevallard, Y. (2017). La TAD et son devenir: Rappels, reprises, avancées. In G. Cirade *et al.* (eds), *Évolutions contemporaines du rapport aux mathématiques et aux autres savoirs à l'école et dans la société* (pp. 27–65). Toulouse, France: Université Toulouse Jean Jaurès.

Chevallard, Y. (2019). On using the ATD: Some clarifications and comments. *Educação Matématica Pesquisa*, 21(4) (1–7).

Chevallard, Y., & Cirade, G. (2009). Pour une formation professionnelle d'université. Éléments d'une problématique de rupture. *Recherche et formation*, 60, 51–62.

Cirade, G. (2006). Devenir professeur de mathématiques : Entre problèmes de la profession et formation en IUFM. Les mathématiques comme problème professionnel (Doctoral dissertation). Université de Provence Aix-Marseille I, France.

Cirade, G. (2008). Les angles alternes-internes: Un problème de la profession. *Petit x*, 76, 5–26.

Cirade, G. (2011). Un passé qui ne passe pas: Le cas des "devoirs à la maison". In M. Bosch *et al.* (eds), *Un panorama de la TAD* (pp. 299–320). Barcelona: Centre de Recerca Matemàtica.

Cirade, G. (2019). Infrastructures didactiques pour la formation des professeurs: Le cas de l'étude de praxéologies d'enseignement. *Educação Matématica Pesquisa*, 21(4) (338–356).

Crumière, A. & Cirade, G. (in press). L'organisation et la gestion de données au cycle 4: Quelles difficultés? In H. Chaachoua *et al.* (eds), *Méthodes de recherche en TAD*.

Etzioni, A. (ed.) (1969). *The semi-professions and their organization: Teachers, nurses, social workers*. New York: The Free Press.

Ministère de l'Éducation Nationale (2013). Référentiel des compétences professionnelles des métiers du professorat et de l'éducation. *Bulletin officiel n° 30*. www.education.gouv.fr/p id285/bulletin_officiel.html?cid_bo=73066.

Pressiat, A. (2017). Éléments sur les apports de la théorie anthropologique du didactique à la profession et leur réception. In G. Cirade *et al.* (eds), *Évolutions contemporaines du rapport aux mathématiques et aux autres savoirs à l'école et dans la société* (pp. 679–725). Toulouse, France: Université Toulouse Jean Jaurès.

11

THE EDUCATION OF PROSPECTIVE EARLY CHILDHOOD TEACHERS WITHIN THE PARADIGM OF QUESTIONING THE WORLD

Francisco Javier García, Tomás Ángel Sierra, Mercedes Hidalgo and Esther Rodríguez

1 Teacher knowledge and teacher education in the ATD

The education of prospective teachers is a critical challenge in any society. Well educated teachers are supposed to use more effective pedagogies that should lead to improved student learning. Despite the considerable progress made by researchers in mathematics education, what teachers should learn and how to best support them in this process is still an open problem. This chapter focuses on how these two challenges, the what and the how, are tackled in the anthropological theory of the didactic (ATD). We will start by introducing some theoretical considerations about teacher knowledge and teacher education in the ATD. Sections 2 and 3 specifically argue about the renovation of the teacher education paradigm that persists in many countries. Section 4 reports on a teacher education process with prospective early childhood education teachers in Spain that attempts to bring this new paradigm to life. This chapter concludes by reflecting on some difficulties and constraints faced when attempting to implement this renewed teacher education process.

Within the ATD, teacher knowledge is modelled in terms of mathematical and didactic praxeologies, which are integrated to form teachers' *praxeological equipment* (see Glossary). Meaningful teacher learning takes place by questioning his/her professional world and by searching for possible answers. These fundamental ideas are developed in this section.

The ATD (Chevallard and Sensevy, 2014) is built around the notion of praxeology as a unifying model of human activities:

> The notion of praxeology is at the heart of the ATD. This notion generalises different common cultural notions—those of knowledge [savoir] and skill [savoir-faire], a word that generically refers to "an ability that has been

acquired by training". It should allow designating any possible structure of knowledge, without epistemological-cultural implications (this is knowledge, this is not, this is "just" a skill, etc.), and without attributing an *a priori* or *a posteriori* value to it.

(Chevallard, 2009, p. 4, our translation)

Researchers have extensively and fruitfully used the praxeological analysis of mathematical activity within teaching and learning processes by revealing the transformations that mathematical knowledge undergoes in its transition between different institutions (the didactic transposition process). It corresponds to the perspective of the didactics of mathematics as the science of the conditions and restrictions that affect the social diffusion of praxeologies (Chevallard, 2009).

Not only can the mathematical activity be described in terms of praxeologies. Teaching activities can also be characterised by considering the types of tasks that teachers face, the techniques they deploy, and the discourses (*technologies* and theories) that implicitly or explicitly explain and justify their teaching *praxis*. In this case we can talk about *didactic praxeologies*. Thus, any teaching and learning process—any *study* process—can be described as the joint activation of mathematical and didactic praxeologies in mutual dependency.

When teachers facilitate a study process, they need to bring into play a combination of mathematical and didactic praxeologies. In fact, what determines teacher actions in the classroom is the combination of the praxeological components they have at their disposal at a given time and the way they activate them with certain conditions and under given constraints (Bosch & Gascón, 2009). The term *praxeological equipment* (Cirade, 2006) has been coined to denote this complex combination of mathematical and didactic praxeologies that makes up teacher knowledge. This equipment should not be considered static, but something that is continuously expanding and evolving.

Besides, Chevallard (2009) emphasises the collective and institutional nature of this equipment by hypothesising that teachers' praxeologies do not depend primarily on teachers as individuals, but mainly on the institutions where they act. Thus Chevallard (2007) postulates a dialectic relation between persons and institutions: "persons are the makers of institutions which in turn are the makers of persons" (*ibid.*, p. 132). He points out that persons normally get into already existing institutions and, by becoming their "subjects", they come into contact with an already existing institutional equipment and contribute to making it exist and evolve.

The social world is replete with "institutional idiosyncrasies", which are the main determinants of an individual's behaviour, on which personal idiosyncrasies operate as second-order corrections.

(Chevallard, 2007, p. 132)

In summary, research in the ATD adopts a different perspective to tackle the teacher knowledge problem. First, because it hypothesises the institutional character of this knowledge. Second, because it explicitly questions how mathematical knowledge is conceived and structured within a given institution. Finally, because it integrates teachers' knowledge (*logos*) and teachers' actions (*praxis*) in a unifying model (praxeological equipment), unlike other theoretical approaches (e.g. Venkat & Adler, 2014) that focus mainly on the *logos* (teachers' knowledge).

This alternative formulation has some advantages. It makes it possible to bring into question the problems that teachers face and the techniques they can activate to tackle them: how are these techniques defined?; what kind of techniques exist?; where do they come from?; how can they be developed, evaluated, improved, eliminated, etc.? Besides, it is also possible to challenge the discourses used within the teaching profession to explain and justify certain teaching practices (where do these discourses come from?; how do they evolve?; how do they affect teaching practices?). Finally, this also enables an analysis of the connections or disconnections between practical and theoretical blocks. For instance, one can ask if didactic praxeologies can only evolve from the natural development of techniques and spontaneous technological-theoretical discourses or if there is some kind of systematic preparation of teaching practices' discourses necessary to develop the teaching profession. We can also ask how already existing technological-theoretical developments derived from research into mathematics education impact teaching tasks and techniques and what is the impact of those that come from more general disciplines, like pedagogy or educational psychology (Bosch & Gascón, 2009).

From this perspective, we propose reformulating the research problem of teacher education as two intertwined problems. On the one hand, the characterisation of the praxeological equipment needed, or that is at least useful, for teachers to intervene effectively and pertinently in their students' mathematical education in a given institution. On the other hand, the problem of the instructional formats—or educational *devices*—that can be used to support teachers to gain access to this equipment.

In this reformulation of the teacher education problem, the terms "effectively" and "pertinently" remain unclear. To be meaningful, it would be necessary to define what is assumed to be effective and pertinent teaching of mathematics. Although "effective teaching" is a topic that research has addressed, here we prefer to adopt the perspective introduced by Gascón and Nicolás (chapter 1) about the normative character of didactics. According to these authors, didactics of mathematics, like other social sciences, cannot make value judgements about the teaching of mathematics. The aim of didactics of mathematics, as a scientific discipline, is to question the means that are rationally suitable to achieve certain predetermined goals. Consequently, the problem of teacher education cannot be isolated from the questioning of the envisaged didactic paradigm that determines, to a large extent, the praxeological equipment the teaching profession could need, as well as the *devices* that could grant access to it.

2 Teacher education in the paradigm of questioning the world

The distinction made by Chevallard (2015) between the *paradigm of visiting works* (also called a *monumentalistic* paradigm) and the *paradigm of questioning the world* has already been presented and discussed in part III of this book (chapters 7, 8 and 9). Chevallard (2015) introduced these paradigms mainly in the school context. However, here our position is to use them when applied to the education of teachers. Quite often, teacher education is organised according to the premises of the paradigm of visiting works. This happens when teacher education programmes are organised as a collection of already existing finalised works that a teacher educator "shows" to prospective teachers as worth-knowing pieces of knowledge that are potentially useful for them as future practitioners of the teaching profession (sometimes mathematical works like number systems; sometimes didactic works like Dienes' theory and the use of multi-base blocks).

More often than not, the organisation of teacher education programmes are more based on the kind of *answers* that facilitators (educators) can offer (considered useful to teachers) than on a real *questioning of the world of the teaching profession*. Thus there is a risk of structuring teacher education from an already existing praxeological equipment, normally from their technological-theoretical blocks, by keeping implicit or even missing two main aspects. On the one hand, their raison d'être (rationale)—that is, the professional questions that gave rise to the creation of this equipment. On the other hand, the praxis that these technological-theoretical blocks could give rise to. Without their rationale and their attached praxis, this equipment is transferred with few opportunities for teachers to make sense of it and use it in real school settings. Thus, it is left to the teachers' responsibility how to make this equipment work in specific professional situations (Bosch & Gascon, 2009).

On the contrary, a fundamental anthropological hypothesis is that meaningful learning takes place when it emerges from exploring problematic and genuine questions. This has been captured within the *paradigm of questioning the world*, which we transfer here from mathematics education to teacher education. According to Marianna Bosch and Josep Gascón (2009), teacher education should start from those professional questions that are considered crucial for the teaching profession (or at least potentially crucial). These questions could include personal needs (the individual dimension), but they should go beyond them by becoming questions of the teaching profession (the institutional dimension).

To be more precise, Chevallard (2009) introduced the Herbartian schema (Figure 11.1) to model a didactic system that was coherent with this paradigm. Reinterpreted in the case of teacher education, the schema represents a didactic system formed by a group of teacher-students X, supported by one or several

$$[S(Y, X, Q) \rightsquigarrow \{ A_1^\Diamond, A_2^\Diamond, ..., A_m^\Diamond, W_{m+1}, W_{m+2}, ..., W_n \}] \rightsquigarrow A^\heartsuit.$$

FIGURE 11.1 A model of a study process according to the *questioning of the world paradigm*

"teacher educators or facilitators" Y (with eventually, $Y = \emptyset$), and articulated around the study of a professional question Q (or several questions). The in-depth study of Q initiates an inquiry process known as a *study and research path* (SRP) (Chevallard, 2015) within the teacher education framework, in our case a *study and research path for teacher education* (SRP-TE). The study community takes the study of question Q in-depth. Thus, the community enquires within the world of already existing answers $(A_i^\Diamond)$ by also considering already existing works (W_j) that could be helpful to study, question, deconstruct and validate these answers. The study of such works could also entail considering new questions (Q_k). The ensemble of answers $(A_i^\Diamond)$, works (W_j) and derived questions (Q_k) acts as a milieu (in the sense of Brousseau, 2002), and helps the community to come up with a possible own answer $A^\heartsuit$. This answer should always be considered provisional depending on its appropriateness and power to offer a satisfactory answer to Q .

Bosch (2018) distinguishes between finalised and open SRP, depending on whether the question and the inquiry process have been designed to meet some already chosen answers and works (praxeologies) or if it is the study community that decides what answers are worth considering and what works may be useful to their study. This distinction can also apply to SRP-TE.

Considering teacher learning from this perspective entails a significant challenge as the problem has widened. It includes not just the characterisation of the praxeological equipment, and of the effective ways for teachers to gain access to and operationalise it, but also the identification of both the *crucial questions* for the teaching profession and the processes that can help teachers create *satisfactory answers* to these questions. Figure 11.2 synthesises the connections between modelling teachers' knowledge and the different formulations of the teacher-learning problem from the ATD perspective.

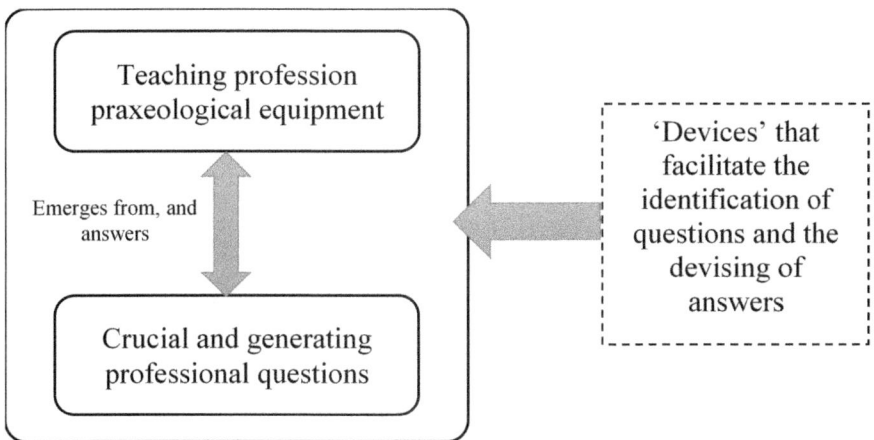

FIGURE 11.2 Reformulation of the teacher education problem in the ATD: teachers' praxeological equipment develops from exploring professional questions, supported by specific education devices

3 The education of prospective early childhood education teachers from the paradigm of questioning the world

This section focuses on the initial education of prospective early childhood education (ECE) teachers in Spain to teach mathematics. The aim is to explore how it could be structured coherently with the paradigm of questioning the world of the teaching profession, as explained above. Therefore, the research problem we are considering can be formulated as follows:

> What are the professional questions to which the education of prospective early childhood education teachers should give answers, and how could these questions be structured to give rise to a training programme?

It is essential to highlight that it is not possible, or even desirable, to offer either a closed list of questions or a definitive complete set of answers. As in many other professions, both questions and answers are open and should be continuously reconsidered. Besides, we should state something about the context: in the Spanish educational system, ECE (with pupils aged 3–6 years) is conceived as an educational stage. Although "playful learning" is central, there is also a syllabus that teachers have to teach. In mathematics, it covers basic logical activities (like classification, ordering, and symbolising), numbers and numbering (in both the cardinal and ordinal meanings), basic arithmetic, and spatial knowledge and shapes.

To identify and formulate possible professional questions from a pragmatic perspective, we decided to start with the big domains in which mathematics is normally structured in the early years. However, this is just one option among others that are open to further discussion. Following the research by Guy Brousseau and colleagues at the Michelet School in Talence (Bordeaux, France), we propose the following domains:

1. Logical activities: symbolisation, classification and ordering, patterns
2. Natural number: cardinal and ordinal numbers. Addition and subtraction
3. Initiation to measurement: length, time, volume, weight and area
4. Spatial and geometrical knowledge: relative positions of objects in space; orientated movements; geometric shapes and solids.

Tomás Sierra, Marianna Bosch and Josep Gascón (2012) have already identified possible mathematical and didactic questions that could be considered typical of the teaching profession in ECE. Here are some questions that might be considered, as well as some possible questions deriving from them.

Exploring any of these questions would exceed the limits of this chapter. Adopting the paradigm of questioning the world implies that exploring any question could lead to both answers and new questions to be explored. Therefore, it is

TABLE 11.1 Possible professional questions for ECE teachers

Question	Derived questions
What is the role of mathematics in our society?	Why should mathematics be studied? What is mathematics for? What does doing mathematics in ECE mean? Why are some activities labelled as mathematical while others are not? …
What mathematics should be taught in ECE?	What are the crucial questions in ECE mathematics that need to be solved? Which of these questions have a strong functional legitimacy in terms of their study facilitating pupils' access to mathematical knowledge in primary education? How to deal with the teaching of mathematics in ECE? What does the globalised teaching of mathematics in ECE mean? …
What kind of mathematical activity could be done in ECE?	What questions confer sense to the mathematical contents taught in ECE? (i.e. what is their raison d'être?) What is the difference between questions that allow the construction of mathematical knowledge from questions that require the use of such knowledge? What is the difference between functional activities, ritual activities, traditional games (e.g. board and cards games) and those designed for pupils to construct specific mathematical knowledge? What is the role that didactic materials (e.g. Cuisenaire rods or Dienes logic blocks) play in the study of mathematics in ECE? How could new technologies be integrated into the study of those questions requiring the use of mathematics in ECE? …

not a matter of merely exploring a single question but also of considering the whole inquiry process that any of these questions could give rise to.

However, we believe it is important to develop at least one case in detail. In the next section we present a teacher education process with pre-service teachers about numbers and its teaching in ECE as a collection of questions and possible answers.

4 A proposal for the mathematical and didactic education of prospective early childhood education teachers

In this section we describe and analyse a teacher education process in pre-service teacher education of early childhood education teachers (ECET), designed as a finalised SRP-TE. The experience took place over several years at the Complutense University of Madrid in Spain from 2009 to 2015. This section is restricted to their initial education as mathematics teachers, bearing in mind that the initial education of ECET is more general and includes other courses of general pedagogy and psychology, sociology, theory and history of education, didactics of social and experimental sciences, etc.

The initial education of ECET at the Complutense University of Madrid spans four years. With mathematics, it comprises two courses: "Developing logical and mathematical thinking and its teaching I" (1st semester of the 2nd year, 45 hours) and "Developing logical and mathematical thinking and its teaching II" (1st semester of the 3rd year, 45 hours). In addition, prospective teachers can take optional courses. In this chapter, we focus on the teachers in the first course, which includes the following topics:

- Topic 1: Didactics of mathematics and early childhood education: the mathematical activity.
- Topic 2: Logical activities in early childhood education.
- Topic 3: Natural numbers in early childhood education.

Topic 1 covers the national and regional ECE curricula, as well as the main elements of Brousseau's theory of didactical situations (Brousseau, 2002). However, in our proposal, this topic was not introduced beforehand using plenary sessions. On the contrary, the main curriculum elements, along with Brousseau's theory, were introduced as they were needed in the process of exploring professional questions, as we go on to see.

The second and third topics deal with some logical and mathematical content and its teaching (for example, *classification* activities and how to teach them in ECE, or natural numbers and how to teach them in ECE).

This distinction between contents and their teaching is especially relevant as it manifests two interconnected challenges:

a   A mathematical engineering challenge: building the components of the mathematical praxeology (MP) at stake.
b   A didactic engineering challenge: designing a didactic praxeology that could make this MP emerge, live and develop adequately in ECE.

As a consequence of this mutual dependency between mathematical and didactic knowledge, prospective teachers will develop their praxeological equipment about these mathematical topics and their teaching in the ECE institution.

The teacher education process that we report here was designed as SRP-TE as part of the *paradigm of questioning the world of the teaching profession*. As with any other study and research path, the core of the process is a set of questions—both the initial one and the questions deriving from it. The open nature of a study and research path means that not all the questions can be determined in advance. Consequently, some of the questions that emerged during our experience can be modified, extended or ignored, and other questions might have been possible, even if they do not appear here.

However, in the context of the pilot experience reported herein, answers to questions were searched mainly in the theory of the didactical situations (TDS) (Brousseau, 2002) and the ATD, which thus became the basic ingredients of the technological and theoretical block of the intended praxeological equipment. Thus, we should talk about a finalised SRP-TE (Bosch, 2018). This was a design decision, connected with ecological constraints, which is not possible to explain here. Considering answers from other technological-theoretical domains, even if the teacher education process starts from the same question, would lead to different developments. Decisions like these could affect both the questions that might emerge from the initial one and the praxeological equipment finally constructed by the prospective teachers involved in the process.

In the next sections we describe the work of a group of prospective teachers, guided by a teacher education facilitator (the second author of this chapter), in which they explored Topic 3: natural numbers in ECE. Specifically, we describe on the one hand the questions that were proposed and how the group tackled them and on the other hand how prospective teachers' praxeological equipment was developed and grew as they looked for answers to these questions. For the sake of clarity, we distinguish between the mathematical dimension and the didactic dimension of their praxeological equipment by bearing in mind that both are interconnected and mutually dependent.

4.1 The SRP-TE about natural numbers: launching the process

The SRP-TE started by exploring the following professional question:

Q_0: What kind of tasks should I design for a functional learning of numbers in ECE?

The group of prospective teachers started exploring this question by analysing the current curriculum and by studying how the mathematical contents are expressed in it. The teacher educator proposed questions like the following:

What domains structure the ECE curriculum? How are the mathematical contents distributed in these domains? What are the aims and contents in each domain? What are the assessment criteria in each domain? Why are the domains in ECE not structured according to basic disciplines (mother tongue, mathematics, science, etc.), as in primary and secondary education? What is missing in the ECE curriculum? What is the numeric knowledge that the ECE curriculum prescribes?

Prospective teachers looked for answers to these questions. To do so, they carefully read and analysed some works W_j (national and regional curricula), and finalised with a whole-group discussion.

To foster the search for possible answers to question Q_0, the teacher educator proposed analysing two teaching situations as possible answers $A_i^\Diamond$ (figures 11.3 and 11.4), based on Sierra, Bosch and Gascón (2012).

BOX 11.1 TEACHING SITUATION 1

Materials: different collections of objects (pencils, balls, little toy figures, etc.). Sheets including drawn collections of objects and a label on which the number of items should be written (see the picture below, on the left, by way of example). Conversely, sheets with a number (figure) indicating the items of a collection that pupils should draw/build/colour (see the picture below, on the right, by way of example).

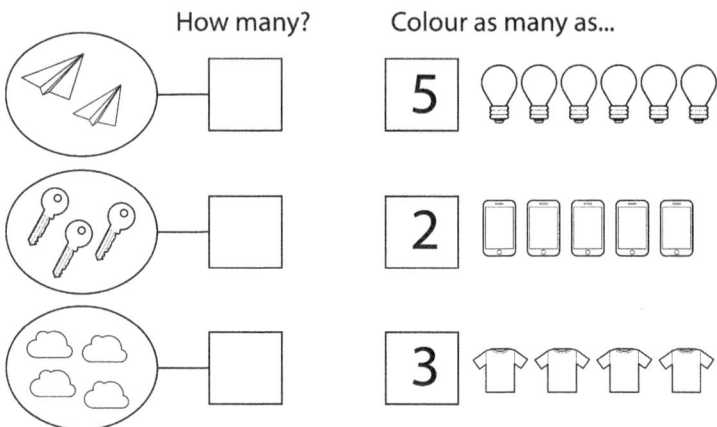

FIGURE 11.3 Teaching situation 1 about numbers in ECE

Development: Pupils sit on the classroom carpet and form a circle. The teacher shows one of these collections (worksheet) and asks: "*How many ... are there?*" Pupils have already seen numbers (figures) up to 6, and today their teacher wishes to introduce the numerals 7 and 8.

To do so, the teacher asks pupils to form collections of seven and eight objects and, conversely, (s)he asks them to state how many objects there are in the collections that (s)he is showing.

After doing some exercises like these, the teacher says: "*Well, today I'm teaching you how to write numbers seven and eight,*" and (s)he writes both numbers on the blackboard.

Pupils are supposed to have learnt the meaning and writing of numbers 7 and 8. Consequently, they keep on doing similar exercises on worksheets to those explained before (given a collection, they have to write the number; given the number, they have to create the collection) with collections of seven and eight objects.

Once pupils can correctly do exercises like these, the teacher moves to number nine (9) following the same teaching plan.

BOX 11.2 TEACHING SITUATION 2: "LAYING THE TABLE"

Materials: a collection of 20 plates, a box with plastic spoons, forks and knives (25 each). Paper and pencil to write messages.

Development: the situation includes several phases:

- 1st phase: pupils work in small groups of four. The teacher places some plates on the table, depending on the aims of the activity and the technique that (s)he wishes pupils to mobilise. (S)He asks pupils to bring the cutlery needed for each plate. In this first situation, the box with the cutlery is placed next to the table with the plates.
- 2nd phase: the situation is the same, but now the box is distant from the table (pupils cannot see the table with the plates when they pick the cutlery). The teacher gives the following instructions to pupils: "You should bring the cutlery so that there is just one spoon (fork, knife) for each plate." One child goes and brings spoons; another brings forks, and so on. In the first stage, pupils can go from the table to the box and return as many times as they wish. But after a while, the teacher introduces a new restriction: "You should bring the cutlery so that each plate has just one piece of cutlery of each type (spoon, fork, knife) but you can go to the box just once." Each pupil goes and puts the pieces of cutlery in a little basket. When they return to the table, the teacher asks: "Do you think you are bringing exactly one piece for each plate?" Next to this pupil, with the help of his or her colleagues, can check if they have correctly solved the situation.

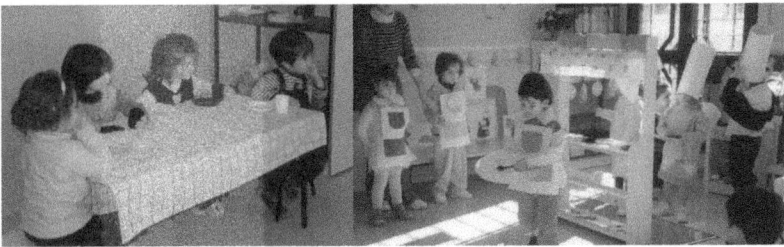

FIGURE 11.4 Teaching situation 2 about numbers in ECE

- 3rd phase: the situation turns into a written communication game. The teacher gives the following instructions to pupils: "Today, you do not pick the cutlery. Instead, you will ask a partner to give you the pieces you need using a written message. Here I will place some plates, and you will write the message for your partner, who can see neither the table nor the plates. (S)He has to give you the pieces you need so that there is just one piece of each type for each plate". The teacher chooses one pupil as the box keeper and another pupil to write the message. Once the pupil brings the pieces of cutlery back, (s)he can check the answer by placing one piece next to each plate. The game is repeated several times, with pupils switching roles.

By analysing these answers, prospective teachers were exposed to two different ways of tackling the learning of natural numbers (in the cardinal sense) in ECE. Prospective teachers' first insights, and their attempts to analyse both situations, made them aware of the limitations of their praxeological equipment. Thus, most of them concluded that the aim of both situations was for ECE pupils to learn to count, even if it was not clear to them what counting actually means (beyond reciting the number words *one, two, three* ... in order) or the relationship between counting and number learning.

Regarding the construction and development of prospective teachers' praxeological equipment, this initial question played an exploratory role. It allowed a first approach to how mathematical knowledge is structured and described in the curriculum (connected mainly with the mathematical dimension of their equipment), but also how it prescribes the teaching of numerical knowledge in ECE (connected mainly with the didactic dimension of their equipment).

The teacher educator aware of the limited development of prospective teachers' praxeological equipment proposed exploring the following question:

Q_1: What does "number" exactly mean in ECE? What does teaching numbers mean (in both the cardinal and ordinal senses) in ECE?

4.2 The SRP-TE about natural numbers: the raison d'être of numbers

Question Q_1 is directly connected with the general problem of the raison d'être of mathematical knowledge in school, which is related with the basic uses and functions of this knowledge (i.e. what this knowledge serves for). The teacher educator attempts to make this connection explicit by formulating the following derived question:

Q_2: What are the types of problems that confer sense to natural numbers in their cardinal sense? What are the questions that need the use of numbers to be answered? What different uses of numbers do we perform? What do numbers serve for in ECE? Is there any previous type of tasks that prepares pupils to learn numbers?

To search for possible answers to these questions, the educator offered different works W_j to prospective teachers: ERMEL (1990), Clements (1999), Pierrard (2002), Martin (2003a, 2003b), Briand, Loubet & Salin (2004), Chamorro (2005), Valentin (2004, 2005), Aguilar *et al.* (2010), Sierra & Rodríguez (2012), Margolinas (2014). Whenever necessary, the educator indicated the chapters or excerpts where relevant information could be found.

Prospective teachers worked in small groups. Some sources were explored during face-to-face sessions in the School of Education, while others were explored outside (autonomous work). After the group work, the teacher educator held a plenary discussion during which the kind of problems that confer meaning to numbers were summarised and institutionalised (and were thus integrated into a possible answer $A^{\heartsuit}$), as well as the possible questions that could give rise to the functional learning of counting and numbers in ECE.

In relation to the development of prospective teachers' praxeological equipment, this inquiry process allowed them to develop its mathematical dimension. They had the opportunity to identify the types of problems that could require the use of numbers (comparing collections, or producing a collection of an equal quantity to a given one, for the cardinal sense; locating an object in an ordered collection for the ordinal sense). Besides, they could also develop the mathematical-didactic dimension of their equipment as they started approaching such situations connected with these types of problems, which could be used in ECE for the functional learning of numbers and numbering (see Lendínez, García & Sierra, 2017).

While looking for answers to previous questions, the notion of *didactic variable* (Brousseau, 2002) implicitly emerged as those aspects in the situation that the teacher could change to provoke and evolution in the MP the pupils are constructing in class.

Besides, the teacher educator used the opportunity to take a step back. That is, to introduce the exploration of those situations that assumedly prepare and help pupils for the later construction of numbers. This led to studying the question:

> Q_3: What are the questions that confer sense to the study of pre-numerical activities? What is the relation between these pre-numerical activities and those that give rise to counting and numbers?

4.3 The SRP-TE about natural numbers: pre-numeric activities

To help prospective teachers to look for a possible answer to question Q_3, the teacher educator offered a work W_j (an excerpt extracted from Sierra & Rodríguez, 2012, p. 29), which included those tasks identified by Joël Briand (1993) as being necessary to correctly perform the counting technique. In order to foster discussion, these tasks were offered as disorganised—thus, prospective teachers needed to think about the order in which tasks should be used in class to lead pupils to need to use counting as the best possible strategy.

The study of the document showed that there are certain task types in which pupils do not need to use number-words, unlike others in which it is completely necessary. The consideration of the former led the group to focus on what Briand (1993) called *enumeration tasks* and to look at the importance of building enumeration strategies with pupils, which are necessary to correctly perform the counting technique later on.

In order to help prospective teachers inquire about enumeration, the following question was posed:

> Q_4: What are the problems or situations that confer meaning to enumeration? That is, what is its raison d'être? What are the possible strategies for solving the enumeration task, and what relationship exists between them? What are the

elements in enumeration situations (didactic variables) that the teacher could change to make pupils develop different strategies to solve them?

Different works W_j were offered to the group to search for answers $A_i^\Diamond$ taken from Briand, Loubet and Salin (2004), Chamorro (2005) and Sierra and Rodríguez (2012).

Regarding the development of prospective teachers' praxeological equipment, the study of these works allowed the identification of *enumeration of a collection* as the mathematical knowledge needed for the later development of counting and, thus, for the learning of numbers. In enumeration tasks there is no need to know how many elements are in the collection. That is why they can be considered pre-numeric tasks. Therefore, prospective teachers could learn about the essential features of an enumeration situation: going through all the elements of a collection, and neither repeating nor forgetting any of them. They could also discover the main techniques that ECE pupils can use to solve them: putting aside already selected elements, marking them, or following any order relation that allows them to go through the collection without repeating or forgetting any element. These elements correspond to the mathematical dimension. In the didactic dimension of their equipment, they could build knowledge about the kind of school situations that could be used in ECE to develop pupils' enumeration strategies. For instance: situations in which children have to perform one action, and just one, on each collection element, but with the condition that a visual control of the action is not possible, e.g. given a collection of opaque piggybanks, inserting just one coin in each piggybank (see Margolinas, 2014).

Furthermore, the teacher educator wished the group to debate the logical activities (designation, classification and ordering) traditionally considered to be necessary for the later construction of numbers. Therefore, sheets from Hughes (1986, pp. 26–40), Brissiaud (1993, pp. 11–24) and ERMEL (1990, pp. 22–28) were distributed as possible works W_j to be studied. These authors consider that designation, classification and ordering activities are essential on their own, but it is not that clear that pupils need to master them before they gain access to numbers (Brissiaud, 1993). Indeed, pupils can start using numbers in situations in which they are the best solution to a given problem, which thus provides them with meaningful and functional learning of them before numbers are mathematically constructed (as an equivalence relation between sets).

4.4 The SRP-TE about natural numbers: mathematical techniques

The genesis and development of mathematical praxeologies are always connected to the development of mathematical techniques (Bosch & Gascón, 1994). Consequently, it is important to question the techniques that could be used to solve the problems that confer sense to numbers, which have been already identified as answers to question Q_2, as well as the connections and evolution that might exist between them. This led to the following question being formulated:

Q_5: What techniques could ECE students use to solve problems that would lead them to count and also to the functional use of numbers? What other techniques could they use to improve the efficacy and economy of counting? What relationship exists between these techniques? Is it possible to sketch a process of their gradual development?

Prospective students revisited the resources (book chapters and papers) that were previously distributed. Their analysis proved useful to identify and name some of the techniques. Later, the teacher educator provided, as an answer $A_i^\lozenge$, a list of possible techniques extracted from (Sierra & Rodríguez, 2012, pp. 28–31), and asked prospective teachers to study and analyse them. This led to a discussion about the mathematical techniques available in ECE that pupils could use to solve the previously described problems. The debate continued to address the problem of a possible hierarchy of these techniques, which could be connected with a possible progression in student learning terms. Ultimately, they faced the problem of the design of a didactic praxeology in ECE that would lead to the gradual development of these techniques to take a certain direction. Once again, the *didactic variable* notion emerged.

During the development of prospective teachers' praxeological equipment, exploring this question allowed expert knowledge about not only counting and how it works but also other techniques—like one-to-one correspondence, one-to-one correspondence using an intermediate collection, subitising, or visual estimation (Lendínez, García & Sierra, 2017)—that pupils could use in ECE depending on the conditions of the situations proposed by the teacher. Besides, more about the didactic dimension of their equipment, they could learn how pupils build these techniques as they face problematic situations with certain characteristics, and how they, as teachers, could make pupils' strategies evolve by modifying some of the characteristics of these situations. For instance, ECET could pose a situation in which pupils had to compare two collections that were separated and not visible simultaneously. Another possible situation consists in producing a collection of an equal quantity to a given one, by asking another pupil for the needed objects orally or in writing, with the extra condition that the initial collection is not visible (Lendínez, García & Sierra, 2017; Margolinas & Wozniak, 2012; Sierra & Rodríguez, 2012).

Having reached this point, and considering that the creation of a didactic praxeology like this is demanding for prospective teachers, the teacher educator formulated the general question Q_6. This question refers to the conditions that a teaching situation should have in ECE so that it integrates the raison d´être of the mathematical praxeologies at stake, and to develop the study process in a given direction.

Q_6: What are the different ways of learning and what are the conditions that a situation should have so that students could make sense of the knowledge that they are learning? What are the features of a study process for it to be considered "functional (offering students the opportunity to understand the meaning and uses of the knowledge they are learning)"?

4.5 The SRP-TE about natural numbers: didactic praxeologies

To look for the answers to question Q_6, prospective teachers started a debate in small groups. The need to analyse, in general, the kind of *mathematical learning situations* that could be used in ECE emerged.

Just as the teacher educator had done previously, a brief literature review offered some elements to build an answer to this question. These included general considerations (suitable for any mathematical praxeology) and other more specific ones, related to the functional teaching of natural numbers in ECE (Sierra, Bosch & Gascón, 2012).

After a complete group debate about the conditions that a learning situation should satisfy, the teacher educator considered that prospective teachers had acquired useful knowledge to go back to the two earlier examples of learning situations (figures 11.3 and 11.4) and carry out a more organised and thoughtful analysis of them. However, thinking that a complete analysis could still be challenging for prospective teachers, he formulated a set of guiding questions that could help them develop a *technique of didactic analysis of teaching situations*. So, first of all, he proposed analysing both situations guided by the following questions:

> Q_7: (a) What kind of problems should be proposed to students?
> (b) What features of a learning situation are characteristic of familiarisation, or of adaptation to the 'milieu'?
> (c) What are the didactic variables? That is, what are the features of the situation that the teacher could change so that it affects students' strategies (their cost, efficacy, complexity)?
> (d) What techniques could students use to solve the tasks proposed in each situation? What is the initial technique that students could use to enter the problem and start solving it? Which techniques are more effective, more economic? Is one of them optimal? What are the connections between the didactic variables and the evolution of techniques? How could the techniques be developed to obtain the optimal one?
> (e) Who validates the solutions to the problems given by students? What ways of validation are available to students?

Prospective teachers had the opportunity to develop and complete their praxeological equipment by integrating into their answer $A^\heartsuit$ the components that they previously built. From both examples they could refine the types of problems that confer numbers and numbering their cardinal sense (similar examples, not included in this chapter, were used to work the ordinal meaning of numbers and numbering). They could also identify the main didactic variables that they could manage to generate an articulated sequence of didactic situations (e.g. situations in which pupils have to produce a collection of equal quantity to a given one: placing collections together, separated but visible, or separated and not visible, accessible or not, etc.), and the strategies that pupils could use. Besides, they could connect the decisions they make about the different didactic variables with the main features of the resulting situation, together with the techniques they wish to promote among pupils and those they want to block. Finally, the contrast between situations 1 and 2 (figures 11.3 and 11.4) allowed them to understand the main characteristics of an "adidactic situation": a situation in which the knowledge to be taught appears as the better tool to solve it, which provokes an interaction between the situation and the pupils and offers feedback to pupils about the appropriateness of their strategy without the teacher intervening. It also shows the power of such situations to provoke improvement in pupils' strategies and favour the functional learning of the mathematical knowledge at stake.

With this study, and by answering all these questions, the teacher educator's aim was not only to develop prospective teachers' competency to carry out this kind of didactic analysis. It also included developing their competencies to design didactic praxeologies that could be used to initiate ECE pupils in the functional use of natural numbers.

5 Conclusions

During the SRP-TE, we found some main constraints that came mainly from the *dominant pedagogical model* that prevails in teacher education institutions and is already described by research (Barquero, 2009). As teacher educators we decided to organise the course around problematic questions instead of using a list of crystallised mathematical and didactic contents. This change led to difficulties in addressing the topics of the course because, for prospective teachers, the answers that could emerge from these questions were not determined beforehand. Sometimes some topics had to be addressed simultaneously, and even then we had to deal with other topics not included in the course syllabus. This kind of teacher education didactic praxeology demands that teacher educators be ready to both consider and analyse new responses and to tackle new questions. In most cases it was not possible to offer a full definitive answer to the questions addressed, which also goes against the usual didactic contract at this educational level. This is somewhat confusing for prospective teachers because, according to the dominant pedagogical model, they assumed that every posed question had an answer and that the teacher educator was responsible for giving this answer.

Moreover, prospective teachers did not have a specific textbook to hand with all the answers they needed to the questions that emerged during the study process. They had to assume new responsibilities that the traditional didactic contract normally assigns to the teacher educator, like posing questions, formulating hypotheses, searching for answers $A_i^\lozenge$ and works W_j, discussing how to build possible answers $A^\heartsuit$, writing and defending reports with partial and provisional answers, and so on. These difficulties, which could, once again, be connected to the dominant pedagogy in teacher education institutions, can only be overcome by designing and implementing new didactic devices that challenge prospective teachers and make them the main actors of the study process.

At first, working in a study community in this way is apparently slower compared to traditional situations in which the teacher educator is the provider of already crystallised knowledge. Managing the didactic time is here an important issue and certainly needs to introduce new teacher education devices. However, teacher educators also need to overcome the *instantaneous learning* requirement, which is typical of the dominant pedagogy (Chevallard, Bosch and Gascón, 1997). It is essential for new teacher education proposals to enable the calm and patient development of activities, including long-term aims. It is also crucial to provide prospective teachers with opportunities to become familiar with the new role they have to play in the didactic relationship, and to get used to a systematic process of searching and analysing the answers they obtain.

In short, a profound change in the dominant pedagogical knowledge is necessary, but also in the epistemological model of mathematics that sustains it (Gascón, 2001). The new model should consider the search for answers to questions as a shared responsibility of the study community (for both prospective teachers and teacher educators).

This kind of change would also need new material and knowledge infrastructures. On the one hand it is important that the furniture in the classrooms allows flexible arrangements to favour group work and discussion, and that the duration of the sessions enables the calm and patient work previously evoked. On the other hand, prospective teachers need to have resources to hand, like books, textbooks, computers, etc., to search for possible answers to the questions they approach. These are not minor aspects for the feasibility of the teacher educational change.

References

Aguilar, B., Ciudad, A., Láinez, M. C., & Tobaruela, A. (2010). *Construir, jugar y compartir: Un enfoque constructivista de las matemáticas en educación infantil*. Jaén, Spain: Enfoques Educativos.

Barquero, B. (2009). Ecología de la modelización matemática en la enseñanza universitaria de las matemáticas (Doctoral dissertation). Universidad Autónoma de Barcelona, Spain.

Bosch, M. (2018). Study and research paths: a model for inquiry. In B. Sirakov, P. N. de Souza & M. Viana (eds), *Proceedings of the International Congress of Mathematicians 2018* (vol. 3, pp. 4001–4022). World Scientific.

Bosch, M., & Gascón, J. (1994). La integración del momento de la técnica en el proceso de estudio de campos de problemas de matemáticas. *Enseñanza de las Ciencias*, 12(3), 314–332.

Bosch, M., & Gascón, J. (2009). Aportación de la teoría antropológica de lo didáctico a la formación del profesorado de matemáticas de secundaria. In M. T. González & J. Murillo (coord.), *Investigación en educación matemática XIII* (pp. 89–113). Santander, Spain: SEIEM.

Briand, J. (1993). L'enumération dans le mesurage des collections: Un dysfonctionnement dans la transposition didactique (Doctoral dissertation). Université Bordeaux I, France.

Briand, J., Loubet, M., & Salin, M. H. (2004). *Apprentissages mathématiques en maternelle* [*Cédérom*]. Paris: Hatier.

Brissiaud, R. (1993). *El aprendizaje del cálculo; Más allá de Piaget y de la teoría de conjuntos.* Madrid: Aprendizaje Visor.

Brousseau, G. (2002). *Theory of didactical situations.* Dordrecht, The Netherlands: Kluwer.

Chamorro, M.C. (2005). *Didáctica de las matemáticas para educación infantil.* Madrid: Pearson Educación.

Chevallard, Y. (2007). Readjusting didactics to a changing epistemology. *European Educational Research Journal,* 6(2), 131–134.

Chevallard, Y. (2009). La TAD face au professeur de mathématiques. Retrieved from http://yves. chevallard.free.fr/spip/spip/IMG/pdf/La_TAD_face_au_professeur_de_mathematiques.pdf.

Chevallard, Y. (2015). Teaching mathematics in tomorrow's society: A case for an oncoming counter paradigm. In S. J. Cho (ed.), *The Proceedings of the 12th International Congress on Mathematical Education* (pp. 173–187). Cham, Switzerland: Springer.

Chevallard, Y., Bosch, M., & Gascón, J. (1997). *Estudiar matemáticas; El eslabón perdido entre la enseñanza y el aprendizaje.* Barcelona, Spain: ICE Universidad de Barcelona & Horsori.

Chevallard, Y., & Sensevy, G. (2014). Anthropological approaches in mathematics education: French perspectives. In S. Lerman (ed.), *Encyclopedia of mathematics education* (pp. 38–43). Dordrecht, The Netherlands: Springer.

Cirade, G. (2006). Devenir professeur de mathématiques : Entre problèmes de la profession et formation en IUFM. Les mathématiques comme problème professionnel (Doctoral dissertation). University Aix-Marseille I, France.

Clements, D. H. (1999). Subitizing: What is it? Why teach it? *Teaching Children Mathematics,* 5, 400–405.

ERMEL (1990). *Apprentissages numériques. (Grande section de maternelle).* Paris: Hatier.

Gascón, J. (2001). Incidencia del modelo epistemológico de las matemáticas sobre las prácticas docentes. *Revista Latinoamericana de Investigación en Matemática Educativa,* 4(2), 129–159.

Hughes, M. (1986). *Los niños y los números. Las dificultades en el aprendizaje de las matemáticas.* Barcelona, Spain: Planeta.

Lendínez, E. M., García, F. J., & Sierra, T. A. (2017). La enseñanza del número en la escuela infantil: Un estudio exploratorio del logos de la profesión. *REDIMAT – Journal of Research in Mathematics Education,* 6(1), 33–55.

Margolinas, C. (2014). ¿Saberes en la escuela infantil? Sí, pero ¿cuáles? *Edma 0–6: Educación Matemática en la Infancia,* 3(1), pp. 1–20.

Margolinas, C., & Wozniak, F. (2012). *Le nombre à l'école maternelle; Une approche didactique.* Brussels: De Boeck.

Martin, F. (2003a). *Apprentissages mathématiques : jeux en maternelle. Livre du maître.* Bordeaux, France: CRDP d'Aquitaine.

Martin, F. (2003b). *Apprentissages mathématiques : jeux en maternelle. Fichier d'illustrations.* Bordeaux, France: CRDP d'Aquitaine.

Pierrard, A. (2002). *Faire des mathématiques à l'école maternelle.* Grenoble, France: CRDP de l'Académie de Grenoble.

Ruiz-Olarría, A. (2015). La formación matemático-didáctica del profesorado de secundaria: De las matemáticas por enseñar a las matemáticas para la enseñanza (Doctoral dissertation). Universidad Autónoma de Madrid, Spain.

Sierra, T. A., Bosch, M., & Gascón, J. (2012). La formación matemático- didáctica del maestro de educación infantil: El caso de enseñar a contar. *Revista de Educación,* 357, 231–256.

Sierra, T. A., & Rodríguez, E. (2012). Una propuesta para la enseñanza del número en la educación infantil. *Números,* 80, 25–52.

Valentin, D. (2004). *Découvrir le monde avec les mathématiques à la maternelle. Cycle 1: Situations pour la petite et la moyenne section.* Paris: Hatier.

Valentin, D. (2005). *Découvrir le monde avec les mathématiques : Situations pour la grande section de maternelle.* Paris: Hatier.

Venkat, H., & Adler, J. (2014). Pedagogical content knowledge in mathematics education. In S. Lerman (ed.), *Encyclopedia of mathematics education* (pp. 477–480). Dordrecht, The Netherlands: Springer.

12

THE EDUCATION OF SCHOOL AND UNIVERSITY TEACHERS WITHIN THE PARADIGM OF QUESTIONING THE WORLD

Berta Barquero, Ignasi Florensa and Alicia Ruiz-Olarría

In this chapter we reflect on what teacher education means within the paradigm of questioning the world. With this in mind, we present a proposal of "study and research paths for teacher education" (SRPs-TE), a teacher education format elaborated within the anthropological theory of the didactic based on the collective study of a professional teaching question carried out by teacher-students under the guidance of educators. After describing the structure of SRPs-TE in five modules, we present three case studies implemented with pre- and in-service teacher-students of different school levels: primary, secondary and university. We point out the potential of SRPs-TE, especially for empowering teachers to carry out epistemological and didactic questioning and for disseminating didactic tools elaborated in research to the teaching practice. We finish by considering the common aspects of the analysed experiences and raising some open issues that might enrich the future development of SRPs-TE.

1 Teacher education within the ATD

Research in mathematics education has been characterised by an important diffusion of inquiry-based study processes during the first two decades of the 21st century (Artigue & Blomhøj, 2013). This phenomenon has its foundations in research work developed during the first half of the 20th century, based on authors such as John Dewey (1938), Jean Piaget (1974) and Georges Pólya (1945). The conceptualisation and description of inquiry-based study processes is mainly done in terms of general notions (such as "active learning", "real world situations", "scientific work", among others) but the specific application and the way researchers, teachers and students have to deal with the knowledge to be taught remain mostly unquestioned.

Yves Chevallard (2015) has described and theorised these changes in mathematics education in terms of a paradigm shift: from the *paradigm of visiting works* to the *paradigm of questioning the world* (see Glossary). Chevallard not only characterises the transformation in mathematics education at the pedagogical level ("how to teach?") but also includes an analysis of the changes that the paradigm shift may have at the didactic level: "what to teach?" In the paradigm of questioning the world, knowledge to be taught is associated with the study of relevant questions. The study of these questions includes moments of study (searching for available answers in the *media*) and moments of research (deconstruction and reconstruction of knowledge to generate one's answer). Implementing question-led study processes prompts the knowledge to be taught to become dynamic, provisional and collective (compared to the traditional notion of knowledge in school institutions).

The profound changes that this paradigm shift causes in the nature of knowledge have two main consequences for teachers. First, it generates new didactic and epistemological needs, not only to design open-ended projects but also to describe, manage and institutionalise the knowledge involved, which is "knowledge in use" and not "crystallised knowledge" (Florensa, Bosch & Gascón, 2015). Second, teachers need to consider knowledge to be taught as something that can (and should) be questioned, which is not a given. Such questioning is not a spontaneous activity for teachers and, consequently, has to be promoted through teacher education.

In parallel to the above, research in teacher education has also evolved over the last 30 years. Shulman (1987) was one of the first researchers to explicitly state that teachers need more than content knowledge (CK) to develop their activity. He proposed three categories of knowledge needed by teachers: CK, pedagogical knowledge (PK) and pedagogical content knowledge (PCK, also named "didactic knowledge" by other authors). PCK is considered a new notion and includes a "wide range of aspects of subject matter knowledge and the teaching of subject matter" (Ball *et al.*, 2008). Ball and colleagues (2008) clarified the PCK notion and systematised its use. These three levels of knowledge needed by teachers are widely received, as one can read in "Solid Findings in Mathematics Education" (Education Committee of the EMS, 2012), which included them as one of the accepted results from research. It is pertinent to highlight that the PCK notion seems compatible with the founding principle of the epistemological approach adopted in this book, including an explicit analysis and questioning of the knowledge to be taught. However, as Carl Winsløw and Viviane Durand-Guerrier (2007) describe, the limits and interrelations of the two approaches remain controversial.

The ATD proposal for teacher education faces two main challenges: first, enabling teachers to deal with questions-led study processes and, second, empowering them to question the knowledge to be taught. Research within the ATD has already shown that, in the case of pre-service teachers, their spontaneous questioning was mainly related to the PK, while PCK and CK remained unquestioned (Durand-Guerrier, Winsløw, & Yoshida, 2010; Winsløw & Durand-Guerrier, 2007). One of the first proposals regarding teacher education within the ATD was the work of Gisèle Cirade (2006). She presents the analysis of more than 7,000 questions raised by pre-service teachers during their internships. Her research highlights that these questions involve didactic aspects of teaching activities and also mathematical elements, asking for the creation of new *praxeologies for teaching*, including the development of new mathematical *praxeologies for the teaching profession* (chapters 10 and 13).

According to these assumptions, Alicia Ruiz-Olarría (2015) and Berta Barquero, Marianna Bosch and Avenilde Romo (2018) present the proposal of *study and research paths for teacher education* (SRPs-TE) as an educational format to provide teachers with pertinent (theoretical and practical) tools to nourish and sustain this type of questioning, making it also operative for the design and management of didactic processes. During the last ten years, diverse experiences of design, implementation and analysis of SRPs-TE have been developed (see Table 12.1) at different school levels and in different conditions (pre-service and in-service teacher education, online or in distance training, etc.). These relevant empirical experiences with teacher education within the ATD lead to the following research questions:

RQ_1:What teacher professional questions have been used as generating questions and how have they evolved during the development of SRPs-TE? Is there a common questioning in the different experiences?
RQ_2: What ATD epistemological and didactic tools have been used by teacher-students in the experiences? How have educators make them available?
RQ_3: What aspects should be enhanced to facilitate future teacher education experiences, taking into account the SRPs-TE institutional conditions and constraints identified in the diverse experiences?

To propose initial answers to these questions, we analyse three case studies of different implemented SRPs-TE. This chapter starts with a description of the generic modules structuring the proposal of SRPs-TE, also including a general overview of the implemented SRPs-TE. Next, we present a more in-depth analysis of three teacher education experiences. The first one is an SRP-TE for primary school teachers, based on a modelling activity about building a box for cakes (adapted from Chappaz & Michon, 2003). The second one corresponds to an SRP-TE regarding functional modelling for secondary school teachers based on a "savings plan" activity (García, Gascon, Ruiz-Higueras & Bosch, 2006). Finally, we present an adaptation of an SRP-TE addressed to lecturers of an engineering school in Barcelona about the teaching of modelling in engineering subjects. We conclude the chapter with the consequences these three case studies reveal.

2 Study and research paths for teacher education (SRPs-TE)

The general structure of an SRP-TE relies on three main features. The first is to define and place a professional teaching question at the starting point of teacher education programmes. Second, educators introduce epistemological and didactic tools as far as they enable teachers to tackle professional problems and provide

TABLE 12.1 SRPs–TE implemented at different school levels

	School level	Type of teacher education	Generating question	Modules 0–4	Previously experienced SRP	Duration
1	Primary	Pre-service	How to show the rationale and functionality of the place-value numeral system?	0, 1, 2	Yes Sierra (2006)	30 hours
2	Primary	Pre-service	How to teach mathematical modelling in primary school education? The cake box activity	All	Yes Chappaz & Michon (2003)	40 hours
3	Primary	Pre-service	How to introduce randomness and inferential statistics in primary school? The balls inside the bottle	All	Yes Brousseau, Brousseau & Warfield (2001)	30 hours
4	Secondary	Pre-service	How to organise the teaching of elementary functional modelling in secondary school education?	0, 1, 2	Yes García (2005)	12 hours
5	Secondary	In-service	How to teach mathematical modelling in secondary school education? The case of forecasting Desigual sales	All	Yes Serrano, Bosch & Gascón (2010, 2013)	80 hours
6	Secondary	In-service	How to teach mathematical modelling in secondary school education? Forecasting Facebook users	All	Yes Barquero, Monreal, Ruíz-Munzón & Serrano (2018)	80 hours
7	Secondary	Pre-service	How to introduce the raison d'être of real numbers in secondary school education?	All	No Licera (2017)	30 hours
8	University	In-service	Evolution of an epidemic Could modelling be the main motivation of my subject?	All	Yes Lucas (2015)	12 hours

answers to the questions faced. Third, besides providing teachers with different design and analytical tools, it is essential to create an empirical *milieu* shared by trainee-teachers and educators-researchers, useful to describe the school constraints affecting study processes. During the SRP-TE, this milieu first includes a mathematical activity, then the design or adaptation of a teaching process fitting with the current school conditions of each trainee-teachers, the implementation of this teaching process and a final analysis. Creating this shared empirical milieu is crucial to "materialise" the specific institutional constraints found by teachers in their daily work, in a way that is objective enough to help in identifying them.

SRPs-TE are structured in five general modules with the following traits.

- *Module 0: Starting from a professional question.* By taking a professional teaching question as a starting point, trainee-teachers ("teachers" hereafter) are invited to search for available answers among the different accessible media (books, textbooks, curricula guidelines, etc.), which eventually include some instructional proposals coming from educational research. The professional question raised in this module will guide the whole SRP-TE, which will finish with the production of a final (provisional) answer to this question.
- *Module 1: Experiencing a study and research path (SRP).* Teachers are asked to act as students within an SRP provided by the teacher educators ("educators" hereafter). The aim of this module is twofold. First, to make teachers carry out an unfamiliar mathematical activity that could, to a certain extent, exist in a regular classroom. Second, to carry out the analysis of the experienced activity with the help of the instructions, using some tools mainly developed in research.
- *Module 2: Collective design of a lesson plan.* Teachers are requested to design an adapted version of the previously experienced mathematical activity for a specific group of students (theirs, if possible). This design takes the form of a lesson plan as close as possible to teachers' practice, including an a priori design of the activity. During this adaptation, it usually happens that teachers "reduce" the potential of the proposed instructional activity to fit in with their school institutional constraints.
- *Module 3: Implementation and analysis of the lesson plan.* Teachers are asked to implement their adapted teaching proposal in a real classroom or with students in an out-of-class activity. In this module, teachers are supposed to use the a priori design as a tool for managing the implementation of the activity and for developing its *in vivo* analysis.
- *Module 4: Collective analysis of the lessons.* This last module is devoted to sharing the teaching experiences (and the constraints faced) with the other teachers under the guidance of the educators. Teachers are asked to share and compare the institutional constraints found and the level at which they manifest themselves. They finally have to present a new adaptation of the instructional proposal and a detailed analysis of the entire process.

The specific activities that are proposed in each module depend on the professional questions and are defined by the educators. Research within the ATD has worked with different kinds of SRP-TE, some going through all the modules, others only focusing on some of them. In many cases, these activities are taken and adapted from previous experienced SRPs. In Table 12.1 we summarise the SRP-TE implemented so far, pointing out some of their characteristics: the school level that teachers are trained to work in; whether participants are in-service or pre-service teachers; the generating questions at the starting point of the SRP-TE; the modules developed; whether the SRP-TE relies on a previously experienced SRP; and the total duration of the implementation.

After this general presentation, we will now focus on three of the implemented SRPs-TE (rows 2, 4 and 8 of Table 12.1). The first experience corresponds to an SRP-TE implemented in a university degree in primary school education. Section 2 describes the content of each module of the SRP-TE, presents an analysis of the modelling activities developed by the teachers during one of the implementations, explains how teachers analyse this activity and describes the use of some didactic and epistemological tools. Section 3 presents an SRP-TE implemented in a master's degree for pre-service secondary school teachers. This section includes the activity developed by the teachers together with its didactic and mathematical analysis. Section 4 includes a description of one of the first instances of an SRP-TE for university teachers. It first presents a general review of the professional development of lecturers at university, marked by a lack of specific teacher education courses. It then describes the experience of adapting an SRP-TE for the lecturers of an engineering school. The chapter concludes with some partial answers to the research questions previously stated.

3 An SRP-TE about teaching modelling for primary school teachers

The teacher education activity about the cake box was designed and implemented with two groups of about fifty pre-service teachers in their 4th year of a university degree in primary school education. It has been implemented over five academic years (from 2014/15 to 2018/19) in the last compulsory course of didactics of mathematics (1st semester, 6 ECTS[1]). The design of the different activities follows the modules of the proposal of SRP-TE with some adaptations for each group of students. Its implementation took 16 two-hour classroom sessions and some more outside classroom students work to implement and analyse a teaching proposal according to the last modules of the SRP-TE.

We can summarise the development of the SRP-TE according to five activities:
First activity (Module 0). Educators raise a general teaching question that is the starting point of the SRP-TE:

> Q_{0_SRP-TE}: How can modelling be introduced in primary school education? What kind of modelling activities could be used?

Teachers are asked to make a first analysis of the role of modelling in mathematics education using different resources, such as the local curriculum, textbooks or some teaching proposals. This activity has a double purpose: first, to describe the conditions under which modelling is proposed to be introduced in primary school education and, second, to establish some shared terms to refer to modelling, such as system, model, variables of the system, mathematical results, interpretation, model validation, etc. At this stage, the educator introduces some research papers (Borromeo-Ferri & Blum, 2010; Wozniak, 2012) to help teacher agree on the modelling words and discourses, most of the time new to the students.

Second activity (Module 1). Educators propose a modelling activity, the *cake box*, which corresponds to an adaptation of the one proposed by Luisa Ruiz-Higueras (2008) as an adaptation of (Chappaz & Michon, 2003). Teachers were asked to assume the role of primary school students and experience the different steps of this activity. It usually takes about three two-hour classroom sessions. The main goal of this module is to make teachers carry out an unfamiliar modelling activity that could, to a certain extent, exist in an ordinary classroom.

Third activity (Module 1). Analysis of the mathematical activity. This activity spanned three more classroom sessions, in which students were supposed to analyse the modelling process as it was experienced by themselves and by other groups in the class.

Fourth activity (Module 2). Teachers had to prepare a lesson plan to implement this cake box activity at primary school level. Then, some of the groups carried out its implementation with primary school students in accordance with their lesson plans.

Fifth activity (Modules 3 and 4). Analysis of the implementation and final answer to the initial question. All the groups shared their teaching experience, and the educator asked them to think about the conditions and constraints relating to the teaching of modelling they had experienced through the different stages of the SRP-TE.

In the next sections we present some of the results from the 2016/17 implementation of this SRP-TE. In particular, we focus on giving some results of the *second activity* when experiencing the cake box modelling activity, based on the reports delivered by teachers and classroom observation. Then we describe the role that the "questions-answers maps" have had in the third and fourth activities when analysing the modelling processes experienced and designing their lesson plan. Finally, we include some of the comments we have collected from groups' final reports, and from the survey distributed at the end of the SRP-TE, where teachers analysed their experience with the epistemological and didactic tools provided.

3.1 Managing with the different modelling phases in the cake box activity under the role of students

When experiencing the cake box activity, teachers succeeded in assuming the role of students and carried out the proposed modelling activity. The activity started by presenting the case of a baker who needs help for packing the cakes in boxes. The baker wants to use the same type of boxes (Figure 12.1). The following question started the activity: "How can we build boxes to help the baker pack the variety of cakes she offers? Which relation exists between the sizes of the initial material (paper or cardboard) and the dimensions of the resulting box?"

From this initial question, the activity is initially structured in three phases depending on what is given and what remains unknown: (1) sizes of the papers; (2) dimensions of the boxes; (3) information about the cakes to pack. We present some examples of questions that have been proposed by the educator (the main ones: Q_1, Q_2, ...) and by the students (the derived ones $Q_{1.3}$, $Q_{3.3}$, ...) working in groups of 3–4 persons.

First phase: We consider that the paper sizes (width and length) are given and focused on question Q_1: *What are the dimensions of a box resulting from a paper whose sizes are fixed?* Students started by considering some particular cases, such as:

$Q_{1.1}$: What dimensions of the box result from A4 paper?

$Q_{1.2}$: If we take an A5 (half A4) sheet, do we obtain a box measuring half of the previous one?

$Q_{1.3}$: What box sizes do we get from a square sheet? Do we get a square-based box?

In this phase, all the students worked building the box and measuring the sizes of the box using different instruments (paper grid, ruler, etc.). Some of the more advanced questions came from the arithmetical comparison of the papers and resulting boxes' sizes: students started formulating initial hypotheses about the likely relationships between the different variables. At the end of this first phase, some of the questions students raised were about a possible relationship of proportionality between the paper and the box sizes, and about how to get a box with a particular size.

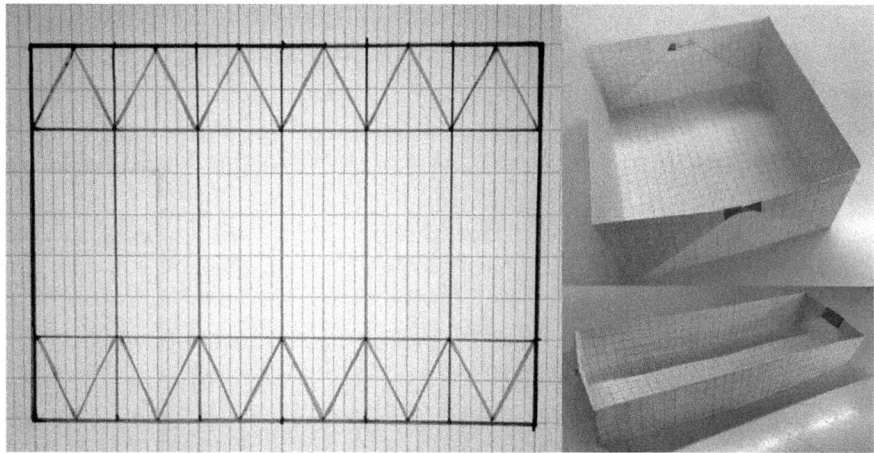

FIGURE 12.1 Instructions to build the box and examples of the resulting boxes

Second phase: We assume that the baker gives us the box sizes she needs. The second question Q_2 is: *Which are the initial paper sizes required to build a box with specific dimensions?* Students proposed derived questions with particular sizes of the boxes, such as:

$Q_{2.1}$: What initial paper sizes do we need to get a box with a base of 6 cm×13 cm? Or, how to get boxes with a square base (such as 5 cm×5 cm or 16 cm², etc.)?

$Q_{2.2}$: Could proportional reasoning be used to find out connections between measurements of the paper and the box?

$Q_{2.3}$: How to modify the paper to get boxes with the same base but of different heights?

Most of the groups proposed to use *models of proportionality* to deduce the sizes of the paper, on many occasions expressed by a rule of three (cross-multiplication). However, they could check that these relations of proportionality did not work (as not all variables maintain these relationships). Students finished by considering *pre-algebraic* and *algebraic* techniques from opening the box and analysing the relationships. This second phase ended when students were able to predict the sizes of the paper/box without manipulating the paper/box.

Third phase: This phase aimed to prepare a written report as an answer to the baker's demand. A list of the cakes the baker wanted to pack, and their sizes were provided. Moreover, students were asked to add a lift-off lid for each box to pack the cakes correctly. This lid is supposed to follow the same pattern of construction and cover the base box packing the cake. At this stage, students had no necessity to build new models, but the main issue was to agree about the way to construct the lift-off lid:

$Q_{3.1}$: How many cms do we have to leave between the box and the lid (margins)?

$Q_{3.2}$: Where to add these differences, to the paper or the box lid? Is it equivalent?

At this stage, teachers were able to formulate the paper–box relationships, and took advantage of using pre-algebraic or algebraic models to find solutions. For instance, in Figure 12.2 there is the explanation of one of the groups that used the relations detected through the graphical representation of the open box (explained pre-algebraically) to find the paper sheet sizes given the size of the box, in this case 17 cm × 17 cm. They explain: "the length of the base of the box multiplied by 3 (17 cm × 3 = 51 cm) is the length of the paper needed; and the width of the base of the box + (2 × half of the length of the base of the box (17 cm + 2 × 8.5 cm) is the width of the paper."

The reports that teachers prepared in groups at the end of this first experience with the cake box activity were a rich milieu for the next steps. They provided a mathematical and didactic analysis of this activity, which was later the base for the

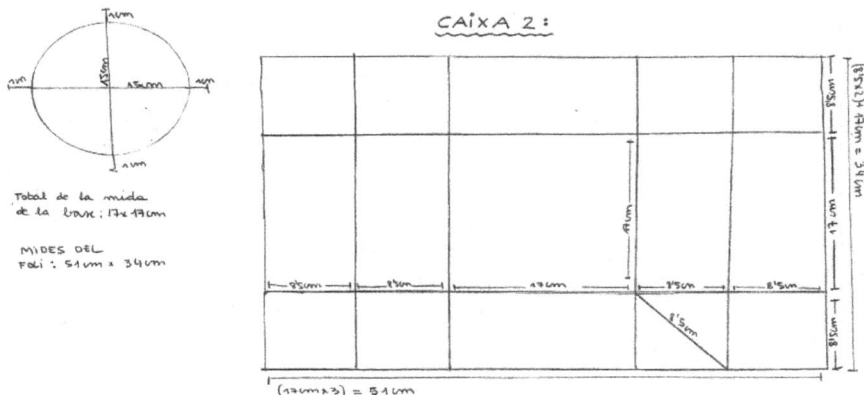

FIGURE 12.2 Pre-algebraic-geometrical models for the box and the lid
(Translation of the handwritten comments: BOX 2: Total measure of the box base: 17 cm × 17 cm. Size of the paper: 51 cm × 34 cm. LID: To build the lid, we add 0.5 cm to each side of the box base to guarantee that it fits correctly. Then, the lid base will be 17.5 cm × 17.5 cm. And, the paper dimension will be 52.5 cm × 35 cm.)

design of their lesson plans. When teachers were asked about the usefulness of this module, they agreed (85.4 per cent of the 105 students) that they were not used to such open activities (56.1 per cent totally agree, 29.3 per cent agree), and that it was an essential step for them, as we can see in these two answers:

> Very useful and not so usual to find in school this kind of activity, also very useful to "put on the skin" of the student to analyse the difficulties they can have.
>
> *(Teacher A)*

> Totally necessary to reflect on the activity from the perspective of the student which let us see the difficulties and challenges that the activity entails.
>
> *(Teacher B)*

3.2 Transferring tools for the epistemological and didactic analysis

In the *third activity*, teachers worked on analysing the modelling work previously carried out, using the session reports and class debates. The educator proposed to do this work using *questions-answers maps* (Barquero et al., 2018). She suggested an initial map with the first questions and answers (Q-A) that appeared during the implementation and asked the students to complete it with the description of their work. This way of describing the modelling process not only provided students with new terminology but also represented an alternative way to talk about doing mathematics, breaking with the usual "static" way of describing school activities, more focused on concepts, notions and techniques to the detriment of questions, models and provisional answers.

Each group worked on producing their Q-A map of the cake box activity, which was used later to analyse the path followed by another group in the class. This task helped them to enrich the Q-A maps by including new questions, answers, strategies, etc., showing the potential of using this tool for future implementation of the activity.

In the *fourth activity*, teachers worked on the design of the *lesson plan* for the implementation of the cake box activity with primary school students. In the class debate, educators agreed, after collecting all the groups' proposals, the elements to include in this lesson plan. In different implementations of the SRP-TE, these elements could change, mainly in the terminology used, but they always belonged to similar categories. Teachers proposed including the Q-A maps. They also proposed, on the one hand, being aware of the mathematical design that supports the proposals and, on the other hand, having "lenses" to analyse students' activity. Then, educators and teachers agreed to include elements related to *topogenetic aspects*—i.e. defining the roles and responsibilities of students and teachers on the different tasks—, *chronogenetic* aspects—i.e. temporary forecast of how to make questions, answers, strategies, etc. evolve over time—, and *mesogenetic* aspects—i.e. different media and means available to students and to the teacher and their role as means for the exploration, validation, evaluation, etc.

When educators asked teachers about this design phase, they reported essential aspects. First, they gave great importance to the mathematical analysis within the lesson plan. When they were asked if "a good a priori mathematical design is important for the lesson guide and its implementation", 78 per cent totally agreed and 19.5 per cent agreed. Moreover, they disagreed with the idea that they could have elaborated the lesson plan without the tools introduced in this course (46.3 per cent totally disagreed, 29.3 per cent disagreed). They highlighted the role of the Q-A maps and the notions of didactic contract as a crucial tool in the lesson plan. In particular, they wrote:

> The questions-answers maps have been a great discovery that, of course, I will consider throughout my professional career, not only in mathematics but in other subjects since it is a tool that allows tracking no content-ended activities and allows a flexible way of managing and evaluating activities […].
>
> *(Teacher A)*

> I believe that it has been very interesting to use such innovative tools for both the mathematical design and for managing and evaluating the teaching proposal. The questions-answers map will be a tool that I will use as a future teacher, since I find that it allows you to develop much more complete activities, to rethink the way of sequencing and of timing them and, also, it is a very good evaluation tool.
>
> *(Teacher B)*

4 An SRP-TE about functional modelling for secondary school teachers

The SRP-TE we introduce in this section focuses on how to teach elementary functional modelling in Spanish compulsory secondary education (12–16 years). It was implemented in the master's degree for pre-service secondary school teacher in the Autonomous University of Barcelona. The design of this SRP-TE followed the same structure of the generic modules described in section 2, but only the first three modules were implemented, as we describe below.

4.1 First activity (Module 0): a professional generating question

The professional question taken as a starting point was formulated as follows:

$Q_{0\text{-}SRP\text{-}TE}$: How to organise the teaching of elementary functional modelling in secondary school education and what is the role of proportionality within this organisation?

Educators presented this question in the first module and teachers had to search, on available resources, elements of response to more specific questions like:

- What is the presence and role of functional modelling in the curriculum of compulsory secondary education?
- Which kinds of functional models are proposed?
- What are these functional models proposed for in the curriculum of compulsory secondary education and for which use?
- What role does proportionality play in the set of elementary functional models that appear in the curriculum of compulsory secondary education?

Teachers, organised in groups of 4–6, looked for answers in available media such as textbooks, teacher training materials, their own experiences as students, research papers, etc. Each group summarised the answers in a written report, before sharing their findings in a whole-class discussion. The main aim of this module was to make teachers problematise the mathematical organisations to be taught around proportionality and functional modelling. Moreover, it aimed to raise the necessity of using this mathematical analysis as the starting point for analysing the associated didactic organisations. In the end it was essential making visible the internal inconsistencies that school mathematical praxeologies presented concerning elementary functional modelling.

4.2 Second activity (Module 1): experiencing an SRP as a student

The purpose of the second activity was to experience a mathematical activity to answer the question: "How to teach elementary functional modelling in secondary education and which role does proportionality has to play?" In this case, we proposed teachers to experience an SRP about *saving plans* as developed in the work of Francisco Javier García (2005) where proportionality is defined as a particular functional relationship between numerical variables among many others. The starting point of the SRP was about looking for the best saving plan for a school class who wants to organise an end-of-course trip. The a priori design of this SRP is similar to the one proposed and implemented by García (2005) with grade 9 students of secondary school (15–16-year-olds). Thus, the generating question of this module, as was presented to teachers, was:

> Q_{0-SRP}: We want to plan the final course trip with enough time. We have to decide 'on different ways to save money to achieve the amount of money needed for this trip. Although we do not know this quantity yet, we can start making an estimation of money needed and taking decisions about our personal savings plans: the number of instalments, shares, etc. It is not our task to decide today how much money we have to give and how, but try to anticipate the necessities we will have when we know the expenses of the trip by the end of the year. What possible savings plans and saving strategies can we consider? What advantages and disadvantages does each one have? How to decide about the terms, the amounts of money in each term, etc.?

The teachers worked in small groups to look for answers to these questions, recording in writing their progress, including details of their discussion, questions, doubts, etc. The role of the educators was limited to making suggestions and raising questions according to their proposals. The next step was that a member of each working group briefly presented the conclusions to the whole class. Then, the entire group was able to work on a shared proposal about the best saving plan to choose and the possible values for the different variables. At this point it was expected that the working groups would raise questions about variables such as the number of instalments and their temporal distribution, the existence of an initial amount and the evolution of the amount given along the savings plan.

The next session was centred on the exploration of new types of savings plan depending on the variables considered and the values assigned to them. Although teachers had a lot of freedom, it was reasonable to expect the emergence of savings plans with equal temporal distribution of instalments and recurrent rules. Following the description provided by García *et al.* (2006, p. 240), the different saving plans prospective teachers were expected to propose could be classified as:

- *Equitable variation savings plans,* in which a fixed amount C is given in each instalment.
- *Accumulative with increasing amount savings plans,* in which a higher amount is given in each instalment. It could be distinguished, depending on the evolution of that amount:

a   If an amount C is given in the first instalment, then the same amount given in the previous one plus C is given in each instalment (if we give C in the first instalment, we will give $C + C$ in the second one, $2C + C$ in the third one, and so on).

b   If an amount C is given in the first instalment then a multiple (by a constant factor $k > 1$) of the previous one is given in each instalment (if we give C in the first instalment, we will give kC in the second one, k^2C in the third one, and so on). For instance, if we consider an increase of 15%, then $k = 1.15$.

- *Accumulative with decreasing amount savings plans,* in which a lower amount is given in each instalment. Depending on the evolution of that amount, it could be distinguished:

a   If an amount C is given in the first instalment then the same amount given in the previous one minus a "discount amount D" is given in each instalment (if we give C in the first instalment, we will give $C - D$ in the second one, $(C - D) - D$ in the third one, and so on).

b   If an amount C is given in the first instalment then a multiple (by a constant factor $0 < k < 1$) of the previous one is given in each instalment (if we give C in the first instalment, we will give kC in the second one, k^2C in the third one, and so on). For instance, if we consider a decrease of 15%, then $k = 0.85$.

During the implementation of this SRP, teachers worked similarly with the different saving plans:

- Characterising saving plans of the different types: equitable and accumulative, using non-complex cases with particular values of the parameters (number of terms n, initial quantity C_0, amount C, increment D, etc.).
- Working with the different saving plans through numerical simulation and analysing these simulations to figure out general conclusions.
- Looking for an *algebraic model* that expresses the final amount saved C_n as a function of the other parameters (number of terms n, initial quantity C_0, amount C, increment D, etc.) of the saving plan (equitable or accumulative).
- Expressing the recursive relation between the final amount saved C_n in each term (n) based on the amount saved C_{n-1} in the previous term ($n - 1$).

During their modelling work, teachers started working with less complicated models about equitable variation saving plans, and they later integrate new hypotheses with different types of variations. Most of them started by using arithmetic techniques with the simulation of the models, manually or using Excel. They readily posed new questions about the limitation of the arithmetical techniques and the necessity of using the algebraic tool to answer some of these questions (see Figure 12.3 illustrating one of the teams' production).

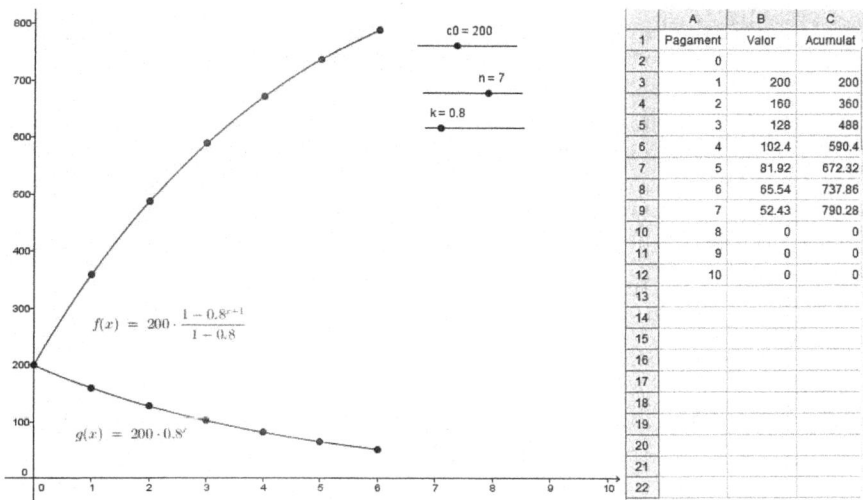

FIGURE 12.3 Work of one of the groups regarding the accumulative with decreasing amount savings plans

This example shows a multiplicative model where every instalment can be obtained by multiplying the previous one by a factor (k) according to:

$$\mathrm{c}n = \mathrm{k c}_{n-1} \Rightarrow \mathrm{c}_n = \mathrm{c}_0 \mathrm{k}^n \Rightarrow \mathrm{S}_n = \frac{1 - k^{n+1}}{1 - k}$$

Figure 12.3 includes the case $n = 7$; $k = 0.8$ and $C_0 = 200$, where function $g(x)$ gives the amount of these instalments, and $f(x)$ the total cumulative amount.

At the end of this module, each group prepared a written summary as an answer to the generating question Q_1. The educators suggested that this summary might include text, tables with the numerical simulations, graphs and formulas. The whole class group proceeded to collect all the answers and to compare and classify the different kinds of saving plans proposed, modelled by different models.

4.3 Third activity (Module 2): analysis of the experienced SRP

This module started with a mathematical analysis of the activity developed in the previous module. Teachers compared the experienced activity to the mathematical school organisations that might reasonably exist at secondary level around elementary functional modelling. Consequently, the questions presented by the educators at the beginning of this module were:

$Q_{1\text{-SRP-TE}}$:: How to describe the mathematical activity developed to generate an answer to the question Q_1? What are the mathematical elements (notions, techniques, properties, etc.) used in the study process?

To clarify these questions, educators introduced some derived questions that might guide the mathematical analysis of the experienced SRP:

> What are the limitations and potentialities of using the arithmetic, algebraic or graphical techniques to study savings plans? During the study of the different saving plans options, have graphical techniques been used? What role has proportionality played in the mathematical activity developed and how has it been related to the other functional relationships?

Each working group proposed an answer to question $Q_{1\text{-SRP-TE}}$ in a written report. Next, the educator gathered all the answers to jointly created a map of the mathematical activity developed in terms of questions and answers. The Q-A map integrated the contributions of all the groups, and each group highlighted the path they followed. This map started from the generating question Q_1 and included the derived questions, the provisional answers to these questions, the hypotheses about the savings plans and the successive questions, answers and assumptions.

This map constituted the first sketch of a mathematical organisation around elementary functional modelling created by the educators and teachers. It included some of the main characteristics of each saving plan, initially through the recursive law that relates each quota with the next one and secondly in terms of the functional models associated to each type of savings plan.

Once the SRP had been described using the Q-A map, the next step was to focus on the analysis of the educators' role and time management. The generating question of this phase of the second module can be formulated as follows:

> $Q_{2\text{-SRP-TE}}$: How has the distribution of didactic responsibilities among the members of the study community? What is the role of the generating question during the SRP? Does the question remain central during the SRP?

To guide the didactic analysis, the educators proposed some derived questions:

> Who is responsible at each moment for deciding the means, the instruments, the adequate techniques to continue the study? How is the type of problems to be studied decided at each moment, as well as the direction that the study process should take at each moment? What criteria are used to decide to deepen in a specific type of problems or, on the contrary, change the activity to study another type of problems? Who is responsible for evaluating the partial results and the provisional answers that appear during the process?

Each working group prepared a written report as an answer to question $Q_{2\text{-SRP-TE}}$ using the derived questions stated. The most relevant reactions of the teachers can be summarised as follows:

- The SRP enables the connection between different functional models and, in particular, the integration of proportionality as a functional relation among others.
- The mathematical activity around functional modelling forces explicit work with parameters and variables. This work illustrates a higher level of algebrisation compared to the usual practice at the secondary level (Bolea, Bosch, & Gascón, 1998).
- This teaching proposal originates a new distribution of responsibilities about the usual didactic contract. In particular, those related to work planning, validation of partial responses and choice of analytical tools.

This module ended with a final task consisting of an adaptation of the SRP for students in grade 9 of secondary education in Spain. Prospective teachers were asked to design a possible SRP for secondary school students, taking into consideration the questions already studied. Teachers were also asked to include the tools that would be necessary to enable students to carry out the designed course. Because of time restrictions, these proposals were neither compared to nor implemented in actual schools.

5 An SRP-TE for lecturer education in engineering degrees

University lecturers' professional development is a particular case of teacher training, with some singularities. We consider that these singularities can be summarised in the following. First, lecturers' activity combines teaching with research. Lecturers are usually researchers in the field where they teach and, consequently, lecturers have a position close to *scholarly knowledge*. Another aspect relates to the constraints regarding the definition of *knowledge to be taught* in higher education. These constraints are much less important at higher education level: lecturers are supposed to define it and curricula are much more generic and play a much less important role in higher education institutions (compared to primary and secondary education). However, this closeness to scholarly knowledge and the capacity of lecturers to modify and create new knowledge to be taught is illusory. In fact, different research work within the ATD framework reveals that knowledge to be taught in higher education is considered as a given and remains unquestioned (Barquero, Bosch, & Gascón, 2013; Bartolomé, Florensa, Bosch, & Gascón, 2019; Florensa, Bosch, Gascón, & Winsløw, 2018).

The second singularity of lecturers' activity is that requirements to access the profession are only related to the research activity: usually, a master's degree and a PhD are enough. In contrast, prospective (and in-service) professional development related to the teaching activity is considered as a complementary training in most countries. The pedagogical and didactic professional development of lecturers is highly dependent on the country: there exist many initial and in-service training courses for teachers in higher education with different characteristics. Paul Trowler and Veronica Bamber (2005) analysed many of these existing courses and observed that they are mainly introductory courses to general pedagogical notions lasting less than a semester. Even if implementing courses of this kind is a general trend in many universities, there are only a few countries in which they are compulsory and proposed by a national agency (Norway, Sri Lanka and Ethiopia) (International Consortium for Educational Development, 2015). A relevant example of the difficulty of implementing a compulsory-based policy is the UK case. Universities UK and GuildHE created the Higher Education Academy (HEA), a publicly funded institution promoting excellence in higher education. The HEA developed the UK Professional Standards Framework for university practitioners: this is a programme describing the competencies and values expected in teaching professionals. This programme, including different learning goals and levels of achievement, was complemented by various compulsory courses proposed by the HEA. However, this programme has very much evolved and is now an option for lecturers, and the HEA itself has merged with the Leadership Foundation and the Equality Challenge Unit to form a new institution (Advance HE) focused on assessing higher education institution on lecturer training.

In parallel, research regarding lecturer education remains a relatively unexplored field. There exists very little literature regarding this subject and the few experiences reported involve only general pedagogical content and do not take into account the nature of the knowledge involved in teaching and learning processes. Let us note that no papers on this field were presented at the European conference CERME9 (or at TWG 14, University Mathematics Education; or at TWGs 18, 19 and 20, Teachers' Knowledge, Practices and Education), or at groups looking at teacher training or university teaching at the last ICME 13, except for a preliminary version of the course presented here (Florensa et al., 2017). The structure of ICME13 Topic Study Groups on teacher education is especially revealing in this respect: there were four groups on teacher education, two (in- and pre-service) centred on the elementary level and two on the secondary level, but none on the higher education level. At a recent conference on mathematics education in North America, only Ellis presented research on teacher assistants training (Ellis, 2014). Regarding the presence of papers in journals about lecturers' education, we have found very little production: only two papers (Guasch, Alvarez, & Espasa, 2010; Postareff, Lindblom-Ylänne & Nevgi, 2007) and the *Handbook on Teaching and Learning in Higher Education* (Fry, Ketteridge & Marshall, 2009).

Considering that the previously described ATD conception of teacher education is also valid at higher education level, we present here the implementation of an SRP-TE in a higher education institution, based on Florensa et al. (2017).

We implemented the course in the Escola Universitària Salesiana de Sarrià (EUSS) in Barcelona (Spain). The EUSS is an engineering school that offers degrees in mechanical, electronic, electrical and management engineering. This engineering school has devoted significant efforts to lecturers' professional development, especially in promoting and facilitating a research-based educational innovation policy. One of the decisions facilitating this option is a four-hour time frame with no teaching for all lecturers on Wednesdays. The course was implemented in this time-frame for three weeks.

The course was structured in six two-hour sessions. Because of this time restriction, the structure of the SRP-TE had to be adapted. We planned the course as follows:

- First session: Explicitly state the professional question Q_0 SRP-TE and briefly present the ATD framework including the notions of *praxeology, Herbartian schema, media-milieu dialectics, topogenesis, mesogenesis* and *chronogenesis* (Barquero & Bosch, 2015). The generating question was stated as: "Could modelling be the main motivation of my subject? Which conditions enable and which constraints hinder this modelling activity?"
- Second and third sessions: A SRP was proposed to be carried out in groups of up to three lecturers. "Taking into account the incidence index of the last nine months of a dengue outbreak, could you forecast the incidence index for the next three months (already known)?" (Figure 12.4).
- Fourth session: Lecturers generated a question-answer map of the experienced SRP. Figure 12.5 presents one of the generated maps.
- Fifth session: Lecturers were invited to create new small groups with the colleagues teaching the same subject. They were asked to design an SRP by choosing a generating question in their field trying to overcome some observed didactic facts such as the absence of raison d'être, the disconnections of topics or the scarcity of the experimental work, among others.
- Sixth session: Lecturers shared some possible teaching proposals, as well as conclusions of the whole course.

In the introduction to the fifth session, lecturers were invited to identify didactic difficulties that they would like to overcome through the new didactic proposal.

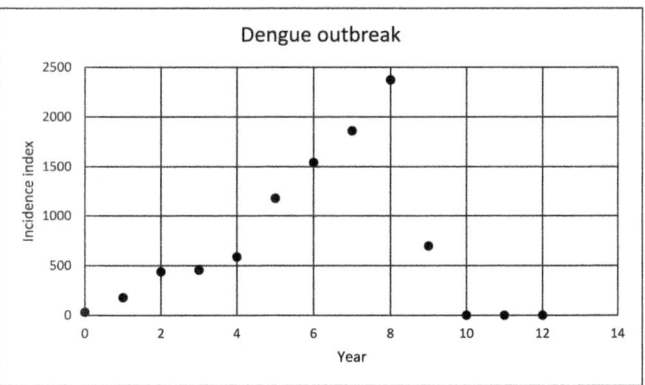

Year	Incidence
0	31
1	179
2	438
3	454
4	587
5	1176
6	1543
7	1859
8	2373
9	696
10	0.1
11	0.05
12	0

FIGURE 12.4 Data used for the experienced SRP

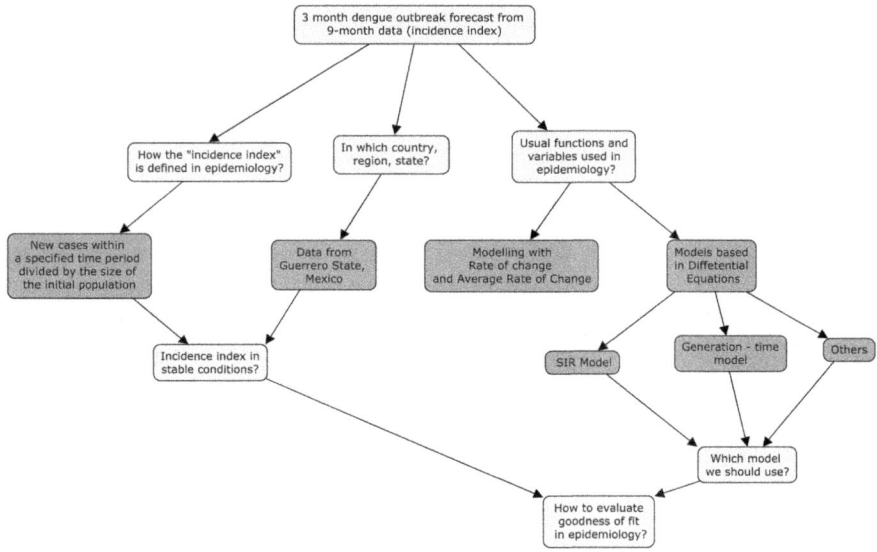

FIGURE 12.5 Question–answer map of one of the groups

The goal was not to implement the inquiry itself but to determine how the dominant epistemology in the institution related to these problematic phenomena and generally propose new possible epistemological and didactic organisations to face them. The question-answer maps were the tool provided to lecturers to carry out this work. During the implementation of the course, some of the content that we initially considered complicated was received more readily than expected (especially the notion of media-milieu dialectics) while some basic concepts proved challenging to share with the teachers—for instance, the description of content in terms of questions instead of topics.

To obtain data to evaluate the course, all the questions-answer maps of all groups, both from the analysis of the actual modelling activity and from the a priori design of the SRP, were collected. We also obtained data from a final survey filled in by all lecturers attending the course. The survey was structured in three main blocks. The first block addressed general aspects of the course such as duration, the balance between individual and teamwork, time structure, etc. The second block asked about the course's content-related issues, such as the work developed with question-answer maps and with the media-milieu dialectics. Finally, the survey asked the lecturers about the possible impact of the course on their teaching activities: changes in the conception of knowledge, dynamics and collective aspects of activities, and the availability of new design and evaluation tools.

The primary outcomes of the implemented course can be divided into research results and consequences for the institution. Regarding the research results, this first SRP-TE addressed to lecturers appears to be a useful tool for empowering lecturers to question and critically examine the dominant epistemology at the university. It produced a discussion (and thus enabled a reflection) on what knowledge has to be taught at the university and how the modelling activity, with its dynamics and collective aspects, could be considered. Regarding the conditions of existence of a lecturer course based on the ATD, it seems that the described conditions make it viable and that some lecturers have taken it as an opportunity to redesign their teaching and learning activities.

Regarding the consequences that the implementation of this course had for the institution, let us highlight that diverse SRPs have been designed and implemented since. First, three editions of an SRP lasting a whole semester in a strength of materials course. Bartolomé *et al.* (2019) present some details of these implementations. Further to this, another three instances of a SRP in a course, Industrial Informatics and Elasticity, can be found in Florensa *et al.* (2018).

6 Conclusions

In this chapter we have described three SRPs-TE implemented under different conditions (pre- and in-service teachers) and at different levels (primary, secondary and university). Also, the length of the experiences varied, ranging from eight to 32 hours. However, common aspects and regularities appear in their analysis.

Regarding the first research question about the professional questions addressed, it is relevant that modelling appears as the central topic in the three cases. In the first experience—the cake box—modelling is explicitly included in the $Q_{0\text{-SRP-TE}}$. In the second experience, it also appears when functions, including proportionality, are conceived as modelling tools. In the third case, modelling appears as a general approach to structure the courses in engineering degrees. This is consistent with the role that ATD assigns to mathematics in the paradigm of questioning the world. In this paradigm, knowledge is valued by its capacity to answer questions through the constructions, use and development of models.

Even if the three experiences were designed independently, we found that the questions addressed in the different activities are quite similar. The first common question was about how to analyse a teaching proposal, and in particular, an SRP. This question was addressed by making teachers experience an SRP in the role of students (as it could exist in a school institution) and then analysing it. The tools provided to do this were also a common aspect of the case studies examined. First, teachers used Q-A maps as an epistemological tool to analyse knowledge developed during the SRP.

Furthermore, this tool played a crucial role when the teachers adapted the experienced SRP in a specific school institution. Here, teachers used Q-A maps not only as an epistemological tool for the a posteriori analysis of teaching proposals. They also used them during the a priori analysis—when they designed the lesson plans and used Q-A maps to anticipate and evaluate possible paths to be followed by students—and in the *in vivo* analysis—when they used them as a tool to institutionalise knowledge.

One of the main weaknesses of these case studies is that there is no systematic way of developing the last generic modules of the SRP-TE, as described in the second section. The impossibility of implementing the adaptations of the SRP in real classroom conditions hinders the discussion about the conditions and constraints experienced in different settings. That is, the ecological analysis of the study process is highly limited, and teachers cannot feel by themselves these limitations. Even if prospective teachers did not implement an SRP during the course, in the first case (primary education) and the third one (university), some teachers have started implementing SRPs in their daily practice since the course finished. In contrast, Barquero, Bosch and Romo (2018) present different experiences of an SRP-TE where this module is fully implemented and this kind of discussions takes place. This research study also shows that the initial modules are crucial to creating the appropriate *milieu* to analyse and discuss the teachers' implementations, a result that is shared with the three case studies considered in this chapter.

Another limitation observed in the three case studies is that educators, and not teachers, assumed the responsibility to raise questions during the development of the SRP-TE, at least the generating and the most important ones. This fact contrasts with the strategy of the "questions of the week" (Cirade, 2006) where prospective teachers raised questions addressed during the course. Therefore, what remains an open question for the development of future SRPs-TE are the didactic devices that may help teachers and educators to pose questions and organise their study jointly. Another open issue is how to exploit teacher education to make the SRP implementations sustainable through the creation of mixed teams with teachers, educators and researchers. This experience has been tested in a preliminary way by Florensa, Bosch, Gascón and Winsløw (2018) with promising results and constitutes a future research line in teacher education.

Note

1 ECTS Stands for European Credit Transfer System. ECTS credits are a standard for comparing the volume of learning for higher education. One academic year corresponds to 60 ECTS credits.

References

Artigue, M., & Blomhøj, M. (2013). Conceptualising inquiry-based education in mathematics. *ZDM Mathematics Education*, 45(6), 797–810.

Ball, D. L., Thames, M. H., Phelps, G., Loewenberg Ball, D., Thames, M. H., & Phelps, G. (2008). Content knowledge for teaching: What makes it special? *Journal of Teacher Education*, 59(5), 389–407.

Barquero, B., & Bosch, M. (2015). Didactic engineering as a research methodology: From fundamental situations to study and research paths. In A. Watson & M. Ohtani (eds), *Task Design in Mathematics Education* (pp. 249–273). Cham, Switzerland: Springer.

Barquero, B., Bosch, M., & Gascón, J. (2013). The ecological dimension in the teaching of mathematical modelling at university. *Recherches en Didactique des Mathématiques*, 33, 307–338.

Barquero, B., Bosch, M., & Romo, A. (2018). Mathematical modelling in teacher education: Dealing with institutional constraints. *ZDM Mathematics Education, 50*(1/2), 31–43.

Barquero, B., Monreal, N., Ruiz-Munzon, N., & Serrano, L. (2018). Linking transmission with inquiry at university level through study and research paths: The case of forecasting Facebook user growth. *International Journal of Research in Undergraduate Mathematics Education*, 4(1), 8–22.

Bartolomé, E., Florensa, I., Bosch, M., & Gascón, J. (2019). A "study and research path" enriching the learning of mechanical engineering. *European Journal of Engineering Education*, 44(3), 330–346.

Bolea, P., Bosch, M., & Gascón, J. (1998). The role of algebraization in the study of a mathematical organization. In I. Schwank (ed.), *Proceedings of the First Conference of the European Society for Research in Mathematics Education* (vol. 2, pp. 135–145). Osnabrück, Germany: CERME.

Borromeo-Ferri, R., & Blum, W. (2010). Insights into teachers' unconscious behaviour in modeling contexts. In R. Lesh, P. Galbraith, C. Haines & A. Hurford (eds), *Modelling students' mathematical modeling competencies* (pp. 423–432). New York, NY: Springer.

Brousseau, G., Brousseau, N., & Warfield, V. (2001). An experiment on the teaching of statistics and probability. *Journal of Mathematical Behavior*, 20(3), 363–411.

Chappaz, J., & Michon, F. (2003). Il était une fois… La boîte du pâtissier. *Grand N*, 72, 19–32.

Chevallard, Y. (2015). Teaching mathematics in tomorrow's society: A case for an oncoming counter paradigm. In S. J. Cho (ed.), *The proceedings of the 12th International Congress on Mathematical Education* (pp. 173–187). Dordrecht, The Netherlands: Springer.

Cirade, G. (2006). Devenir professeur de mathématiques : entre problèmes de la profession et formation en IUFM: Les mathématiques comme problème professionnel (Doctoral dissertation). Université d'Aix-Marseille, France.

Dewey, J. (1938). *Experience & Education*. New York: Macmillan.

Durand-Guerrier, V., Winsløw, C., & Yoshida, H. (2010). A model of mathematics teacher knowledge and a comparative study in Denmark, France and Japan. *Annales de Didactique et de Sciences Cognitives*, 15, 141–166.

Education Committee of the EMS (2012). It is necessary that teachers are mathematically proficient, but is it sufficient? Solid findings in mathematics education on teacher knowledge. *Newsletter of the European Mathematical Society*, 83, 46–50.

Ellis, J. (2014). Preparing future professors: Highlighting the importance of graduate student professional development programs in calculus instruction. In *Proceedings of the 37th Conference of the International Group for the Psychology of Mathematics Education* (vol. 3, pp. 9–16). Kiel, Germany.

Florensa, I., Bosch, M., & Gascón, J. (2015). The epistemological dimension in didactics: Two problematic issues. In K. Krainer & N. Vondrová (eds), *Proceedings of the Ninth Congress of the European Society for Research in Mathematics Education* (pp. 2635–2641). Prague: University of Prague and ERME.

Florensa, I., Bosch, M., Cuadros, J., & Gascón, J. (2017). Teaching didactics to lecturers: A challenging field. In T. Dooley & G. Gueudet (eds), *Proceedings of the 10th Conference of the European Society for Research in Mathematics Education* (pp. 2001–2008). Dublin: Dublin City University and ERME.

Florensa, I., Bosch, M., Gascón, J., & Winsløw, C. (2018). Study and research paths: A new tool for design and management of project-based learning in engineering. *International Journal of Engineering Education*, 34(6), 1–15.

Fry, H., Ketteridge, S., & Marshall, S. (2009). A handbook for teaching and learning in higher education. In H. Fry, S. Ketteridge & S. Marshall (eds), *A handbook for teaching and learning in higher education* (3rd edition). Oxford, UK: Routledge.

García, F. J. (2005). La modelización como herramienta de articulación de la matemática escolar: De la proporcionalidad a las relaciones funcionales (Doctoral dissertation). Universidad de Jaén, Spain.

García, F. J., Gascón, J., Ruiz-Higueras, L., & Bosch, M. (2006). Mathematical modelling as a tool for the connection of school mathematics. *ZDM Mathematics Education*, 38(3), 226–246.

Guasch, T., Alvarez, I., & Espasa, A. (2010). University teacher competencies in a virtual teaching/learning environment: Analysis of a teacher training experience. *Teaching and Teacher Education*, 26(2), 199–206.

International Consortium for Educational Development. (2015). The preparation of niversity teachers internationally (Report). http://icedonline.net/iced-members-area/the-preparation-of-university-teachers-internationally/.

Licera, M. (2017). Economía y ecología de los números reales en la enseñanza secundaria y la formación del profesorado (Doctoral dissertation). Universidad Pontificia de Valparaíso, Chile.

Lucas, C. (2015). Una posible "razón de ser" del cálculo diferencial elemental en el ámbito de la modelización funcional (Doctoral dissertation). Universidad de Vigo, Spain.

Piaget, J. (1974). *La prise de conscience*. Paris: Presses Universitaires de France.

Pólya, G. (1945). *How to solve it*. Garden City, NY: Princeton University Press.

Postareff, L., Lindblom-Ylänne, S., & Nevgi, A. (2007). The effect of pedagogical training on teaching in higher education. *Teaching and Teacher Education*, 23(5), 557–571.

Ruiz-Higueras, L. (2008). Modelización Matemática en la Escuela Primaria: La reconquista escolar de dominios de realidad. In M. M. Hervás (coord.), *Competencia matemática e interpretación de la realidad* (pp. 87–119). Madrid: Ministerio de Educación, Política Social y Deporte.

Ruiz-Olarría, A. (2015). La formación matemático-didáctica del profesorado de secundaria: De las matemáticas por enseñar a las matemáticas para la enseñanza (Doctoral dissertation). Universidad Autónoma de Madrid, Spain.

Serrano, L., Bosch, M., & Gascón, J. (2010). Fitting models to data: The mathematising step in the modelling process. In V. Durand-Guerrier, S. Soury-Lavergne & F. Arzarello (eds), *6th Conference of the European Research on Mathematics Education* (pp. 2186–2195). Lyon: Institut National de Recherche Pédagogique.

Serrano, L., Bosch, M., & Gascón, J. (2013). Recorridos de estudio e investigación en la enseñanza universitaria de ciencias económicas y empresariales. *UNO: Revista de didáctica de las matemáticas*, 62, 39–48.

Shulman, L. S. (1987). Knowledge and teaching: Foundations of the new reform. *Harvard Educational Review*, 57(1), 1–23.

Sierra, T. A. (2006). Lo matemático en el diseño y análisis de organizaciones didácticas los sistemas de numeración y la medida de magnitudes (Doctoral dissertation). Universidad Complutense de Madrid, Spain.

Trowler, P., & Bamber, R. (2005). Compulsory higher education teacher training: Joined-up policies, institutional architectures and enhancement cultures. *International Journal for Academic Development*, 10(2), 79–93.

Winsløw, C., & Durand-Guerrier, V. (2007). Education of lower secondary mathematics teachers in Denmark and France: A comparative study of characteristics of the systems and their products. *Nordic Studies in Mathematics Education*, 12(2), 5–32.

Wozniak, F. (2012). Des professeurs des écoles face à un problème de modélisation: Une question d'équipement praxéologique. *Recherches en Didactique des Mathématiques*, 32(1), 7–55.

13

ON THE CONTRIBUTIONS OF THE ATD TO THE TEACHING PROFESSION

Klaus Rasmussen, Kaj Østergaard and André Pressiat

1 The teaching profession and its problems

In our societies, it is usual to hear slogans like "teachers are key actors in the education of students", "teachers are fundamental" and "we need the best teachers to educate the next generations". Initially, there is nothing wrong with them and, indeed, politicians, journalists, stakeholders, or parents' associations often use this kind of slogan. In the world of education, and particularly in the world of educational research, these ideas are widely accepted without any further questioning.

It is needless to say that teachers play a crucial role in education, but the didactic transposition theory (Chevallard, 1985) revealed that teaching and learning are largely affected by processes that go beyond the classroom and the interactions that take place within it. The development of the ATD, both as a theory and as a research field, has led to a shift from "teachers" as individuals to "teachers" as a collective. Behind this change of perspective lies the hypothesis that teachers should be considered not in isolation but as parts of institutions, and that the relations they establish with mathematics and its teaching can largely be explained by the conditions and constraints that emerge from their links to these institutions, beyond personal idiosyncrasies. Consequently, and as has been pointed out in chapter 10 (Cirade & Crumière), the main focus of the current research within the ATD is not teachers but the teaching profession and its specific needs.

Teaching, as a profession, is subject to tensions coming from many different places. The *scale of levels of didactic codeterminacy* (see Glossary) has proved to be useful to clarify some of these places, being a powerful tool for researchers to enlarge the focus of their research beyond the teaching and learning processes taking place within the classroom. This chapter aims to present some pieces of research that have dealt with problems of the profession connected with different levels of the scale.

Section 2 outlines research related to the mathematical equipment of the teaching profession in France, connected with needs arising from the lower levels of the scale of didactic codeterminacy (discipline and below). It is now well assumed in mathematics education that professional knowledge cannot be restricted to those praxeologies that teachers have to teach. Indeed, the social diffusion of mathematical praxeologies in teaching institutions needs praxeological equipment that goes far beyond them, and that Gisèle Cirade (2006) named *praxeologies for teaching*. We will see some examples of missing knowledge in the praxeological equipment of the profession that could help teachers better understand and plan the teaching of some geometrical objects in the context of secondary education in France.

Section 3 outlines research related to current demands on the teaching profession in countries that plan and implement interdisciplinary teaching, like Denmark. This problem can be placed at the discipline(s) level, although it is affected by conditions and constraints coming from levels above it. The label "STEM education" (Science, Technology, Engineering and Mathematics) had its origin in the National Science Foundation in the 1990s (Bybee, 2010) and, since then, has been spreading quickly in the educational systems of many countries. Whatever it means and entails, it appears as a central challenge for the teaching profession, since it overcomes the traditional division of school disciplines and the way they are taught. The teaching profession needs to develop, even rebuild, its praxeological equipment, in ways that research is still trying to envisage. This section reports on a teacher education programme in Denmark, inquiring about possible ways of developing the praxeological equipment of the teaching profession to teach STEM.

Finally, section 4 goes higher up the scale of levels of didactic codeterminacy (pedagogies and schools) to approach the issue of the collaboration between in-service and prospective teachers. The tension between theory-oriented knowledge, usually built in teacher education institutions, and practical and implicit knowledge, built in day-to-day teaching practices, is addressed as a problem of the teaching profession and, in particular, as a problem of those who are constructing their praxeological equipment to get access to the profession. This final section outlines research conducted in Denmark fostering collaboration between in-service and prospective teachers, which resulted in mutual benefit for both collectives in the development and articulation of the logos and praxis of their praxeological equipment.

The chapter concludes with some implications of the cases reported, which may inspire further research about the teaching profession and the problems it faces.

2 *Praxeologies for teaching* as a problem of the teaching profession

Let us start this section with some facts that could be considered as indicators of problems of the teaching profession regarding their praxeological equipment and, particularly, their *praxeologies for teaching*. The activity of a teacher-researcher working for the education of secondary school teachers is highlighted by fleeting moments during which surprising facts relating to the profession are brought to light. Here are three of them, which the third author experienced in very different situations.

2.1 Moment 1

In a visit during a lesson about expansion-reduction of a figure (grade 8), a teacher-student used a digital photo and several transformations she made on it. One of them was the (orthogonal) stretch map of factor 0.7, the axis of which is the left edge of the photo—she pulled the right edge of the figure to the left. During the after-lesson briefing, with the teacher-student and her teaching advisor, a secondary school teacher with experience, the observer questioned about the name of such a transformation. The teacher-student said she did not know, or no longer knew. When the observer provided the name of "stretch map" with the characteristics (axis, factor) of the stretch, the teaching advisor, astonished, asked: "How did you find this?"

2.2 Moment 2

Figure 13.1 presents an item from PISA 2003 entitled "Bricks".

In a meeting at the Ministry of Education, the third author was asked to comment on the results on PISA for France (which were quite bad). He evoked the originality of the parallel axonometric views used in the item presentation. These views— based solely on what is called "cavalier perspective", with dotted lines for hidden edges (we will explain this notion later)—are not usual in the French educational practice. This comment was not well understood by the audience, which appeared to ignore the meaning of "axonometry"—merely a parallel projection, but with different choices to those made for a cavalier perspective.

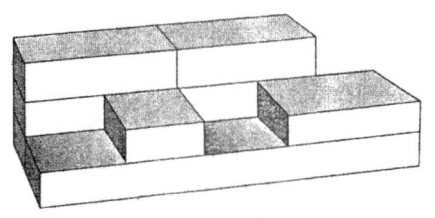
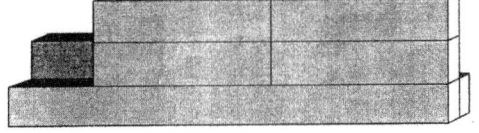

In a stack of bricks, there are three different sizes of brick. Two medium-sized bricks and a small brick put end-to-end, equal to a large brick. Two small bricks laid end-to-end equal one medium-sized brick. Here are two views of a construction made with these bricks. Suppose you want to do the same construction with only small bricks.

How many small bricks will you need?

FIGURE 13.1 PISA 2003 item: "Bricks"

Let us clarify things by borrowing some elements of the *Atlas des Mathématiques* from Reinhardt & Soeder (1999).

Following Figure 13.2 notations, cavalier perspective characteristics are given by the relations $\beta = 0$ and $e_2 = e_3$; military perspective by the relations $\alpha + \beta = 90°$ and $e_1 = e_2$. The original character of the views chosen for the item "Bricks" corresponds to $\alpha = 68°$, $\beta = 6°$ for the first; $\alpha = -1°$, $\beta = 26°$ for the second. In France, the cavalier perspective is considered well known but not mathematised officially. It is still the subject of a legend that ensures that the ratio k of e_1 to e_2 in such a perspective is the cosine of angle α. In everyday life, each of us has been able to observe that the shadow on a wall of an object perpendicular to this wall may be longer than this object: thus, the ratio k can be greater than 1, and this suffices to destroy the legend. But we must distinguish the mathematical cavalier perspective from the practical one and evoke the conventional choices. In practical drawings, in order not to shock the eye, the number k is chosen less than 1: this fact justifies the name "reduction ratio" given to this number. The most frequent value for k is ½, a value chosen when α is 60°, but also when α is 30° or 45°. The relation "$k = \cos \alpha$" of the legend finds a second life in an official document on the Ministry of Education website (MEN, 2012, p. 4): the authors declare that one applies a reduction ratio, "often equal" to the cosine of angle α; on the figure illustrating the cavalier perspective of a cube, $\alpha = 45°$ is chosen, and the reduction ratio is 22 (Figure 13.3). Furthermore, on this figure appears a "horizon line", a line that does not exist in a cavalier perspective. We will talk more about this later.

2.3 Moment 3

To illustrate the plane transformation named "shear", it has become classic to cite its use in simulating italics in text formatting. Letters are often built with Bézier curves and, to modify them, one can simply transform their control

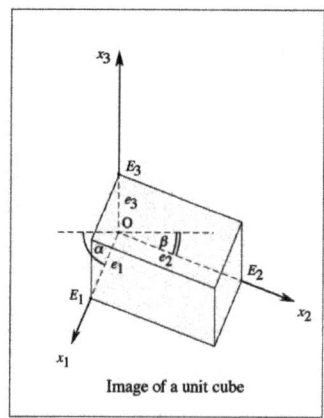

Image of a unit cube

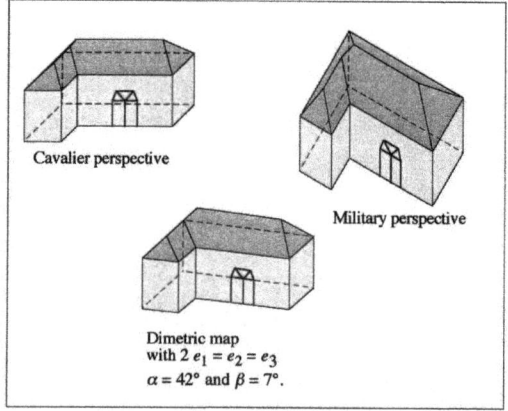

Cavalier perspective

Military perspective

Dimetric map
with $2 e_1 = e_2 = e_3$
$\alpha = 42°$ and $\beta = 7°$.

FIGURE 13.2 Oblique parallel projection
Note: Figures redrawn by the authors, based on Reinhardt and Soeder (1999, p. 176).

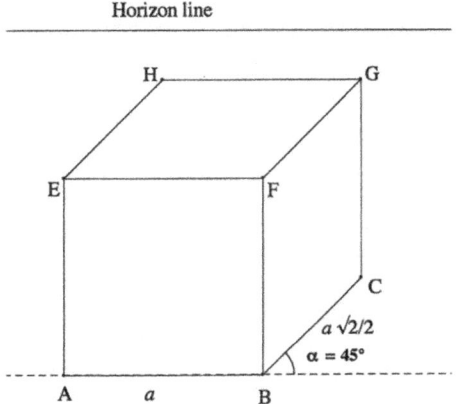

FIGURE 13.3 Cavalier perspective of a cube with α = 45°

points. A rather important development of this issue appears in the Bézier curves chapter of a French recruitment contest preparation book (Mercier, 2012). The transformation used to act on the control points is the subject of a footnote, in which the author gives the name of this transformation: "*On dit que l'on applique un cisaillement, qui correspond à une affinité en mathématiques.*" If we translate this into English, the sentence may sound odd: "One says that a shear is applied, which corresponds to a stretch in mathematics." There is, of course, a confusion between the two geometrical transformations respectively named "stretch"/ "shear" in English and "affinité"/"transvection" in French. A shear is not the same as a stretch: a stretch acting on a line of text would change its length and make it difficult to read.

These three anecdotes concern mathematical objects of geometry and its teaching (parallel projection on a plane, stretch and shear in the plane, cavalier perspective, parallel—or cylindrical—perspective, central—or conical—perspective), about which the profession encounters difficulties of varying importance but the existence of which is indisputable. What mathematical and didactic praxeologies are indispensable or simply useful to shed light on the raisons d'être and advantages of affine geometry? Or to strengthen the ties between affine geometry and parallel projection and so with cavalier perspective? The next sections develop the relevance of the problem raised and describe the inquiry process followed from the perspective of an ATD researcher.

2.1 Relevance of the problem for the profession

The first and third anecdotes question the initial training of mathematics teachers in secondary education in France. They show the difficulty faced by some of its members, at various levels, in transitioning from theoretical geometry to practical or experimental geometry. Following Chevallard (1998, p. 3), we can locate the reconstruction of a mathematical praxeology in a four-dimensional space with a *practical* dimension, an *experimental* dimension, a *theoretical* dimension, and a *formal or mathematical* dimension in the narrow sense. We can assign to these qualifiers the same meaning they have, more conventionally, when we speak of practical physics, experimental physics, theoretical physics and mathematical physics.

This difficulty can be explained—in large part—by the mathematical training received at university, at least in France. At university level, the experimental dimension of mathematical organisations is often forgotten or considered illegitimate; and the training received is hardly more adequate regarding the theoretical dimension. Students are presented theoretically impeccable organisations without participating in their elaboration. Therefore, future teachers become used to high demands in terms of theoretical "finishing" but fail to acquire the capacity to carry out a theoretical elaboration of the mathematical organisations they have to teach. These characteristics of French university mathematical style are obvious when comparing Wikipedia pages written by French scholars with those written by authors from other countries. An example of "shear mappings" and "transvection" is developed in Pressiat (2017) and a brief illustration concerning shears will be given below. Are there other mathematical training praxeologies which better support these needs of teachers about stretches, shears and their links with cavalier perspective? Where and how to find them?

The difficulties that have just been mentioned could be labelled as having "professional" and "societal" reasons. The second anecdote leads us to consider another type of difficulty, one with a curricular reason and specific to French society. The teaching of parallel perspective has been reduced for several decades to that of "cavalier perspective". Why? What are the effects of this reduction on professional praxeologies?

2.2. Inquiry about the problem: selected elements

The inquiry began by the study of the question of the raison d'être of affine plane geometry. It started with Lebesgue's courses about projective geometry in which the author highlights two of Desargues's ideas: the work of artisans as a source of profitable study for science; the interest, from a theoretical point of view, of dealing with the perspective of a plane figure. This led to the geometrical work of Deltheil & Caire (1944): following Lebesgue's second idea, they introduced affine geometry as the geometry of properties of plane figures conserved by parallel projection. In the study of transformations between two projections on the same plane of the same plane figure, stretches appear, with an unnamed transformation, which will be proven to be a shear. Furthermore, any stretch in the plane may be decomposed into the product of a space rotation and a parallel projection.

Brannan, Esplen and Gray (1999) and Yaglom & Shenitzer (1973) follow the same ideas about plane figures and projections, but in more pleasant and less

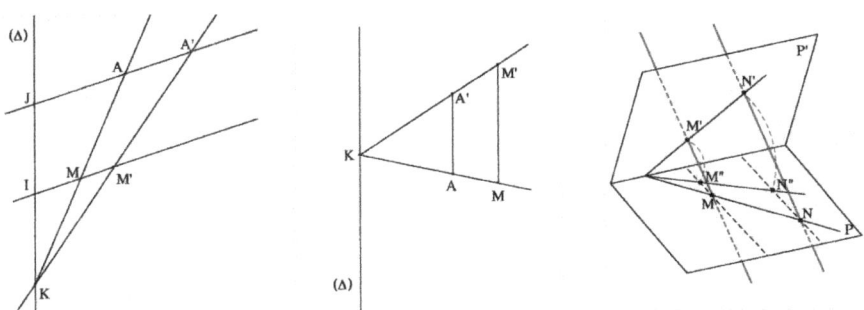

FIGURE 13.4 Parallel projections
Note: Figures drawn by the authors, based on Deltheil and Caire (1944, pp. 270–271) and Reinhardt and Soeder (1999, p. 160).

theoretical ways: both evoke shadows in the sun or rays of light; they avoid the recourse to projective geometry. Yaglom & Shenitzer (1973) use similarities to link affine transformations with "parallel projections of a plane onto itself", which are defined simply. The previous mixtures of parallel projections, rotations of space, or plane similitudes would nowadays sound strange for a French teacher.

An important break, announcing modern maths programmes, was introduced by Georges Bouligand. Following Hermann Weyl views in his famous "Time, Space, Matter" (Weyl, 1952), he proposes a new geometry, named "linear or affine geometry" (Bouligand, 1951), in which the definition of the more interesting transformations no longer refers to space geometry or parallel projections: they are those sending parallelograms onto parallelograms. This idea was used in a didactic way in some German textbooks during the modern maths period (for instance, in Köhler, Höwelmann, & Krämer, 1973; Figure 13.5).

In France, during the modern maths period, such an illustration would not have been accepted. The reason lies in the division of the domain of geometry according to the following sectors, imposed by the programmes: vector spaces,

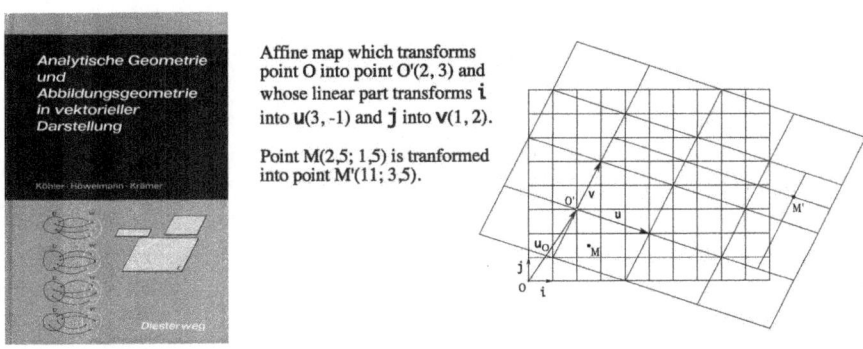

FIGURE 13.5 Affine map transforming a parallelogram into a parallelogram
Note: Figure redrawn by the authors from Köhler et al. (1973).

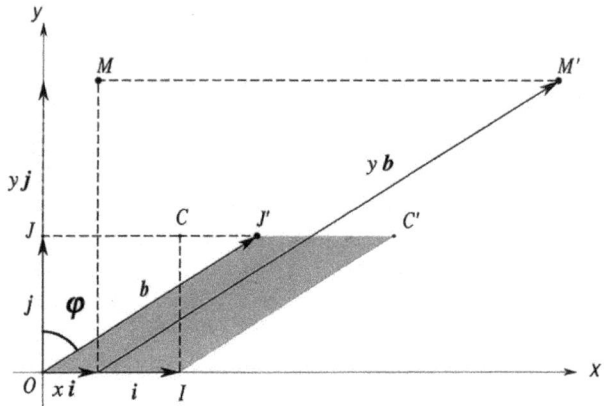

FIGURE 13.6 Transvections, two documents

affine spaces, Euclidean vector spaces, Euclidean affine spaces. In affine geometry, squares do not exist! Projections on a plane in 3D geometry had space for domain and codomain and were not one-one (except the identity map), and so were not a transformation, and parallel projections from a plane onto a plane were out of the programme. Therefore, praxeologies in Yaglom's style were undesirable, while exercises about analytic expressions of affine maps were a major pre-occupation … . Afterwards, projections of all types gradually disappeared (except orthogonal projection) from the mathematical programmes of secondary school. The above division into sectors—vector spaces, affine spaces, Euclidean vector spaces, Euclidean affine spaces—still structures the mathematical university development of teachers.

Let us now come back to shear mappings ("transvections" in French). In Figure 13.6 there are two documents about them. On the left we see the definition of a shear map one can find in a lot of courses at university level in France. On the right, the diagram redrawn by the authors from Köhler *et al.* (1973) provides an axis Ox_1 shear and shows how to define its angle φ. It is easy to imagine the difficulty a student teacher might have endeavouring to link the purely algebraic definition on the left with the geometrical illustration of a shear map on the right … .

With regard to the cavalier perspective, the difficulty for French teachers is related to the need to evoke parallel projection, a content that has progressively disappeared from the programmes over several decades. The consequence is that teachers do not even refer to "shadows in the sun" and reduce the cavalier perspective to a drawing algorithmic technique, this being assumed by a large part of the profession—with some exceptions, such as Bernard Parzysz (1989). An experiment in a class of grade 6 in France, with suitable enriched Zometool equipment, was organised to allow pupils to link parallel projection and the cavalier perspective of a parallelepiped (Figure 13.7, left and centre). The question posed to them was

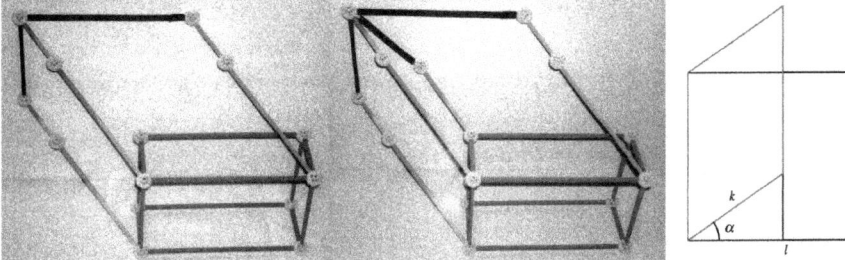

FIGURE 13.7 Cavalier perspective: an experiment in a grade 6 class

"What figure is drawn on the wall if we do the same for each edge of the parallelepiped?"

This hardware device allows showing the parallelepiped, its cavalier perspective and the perspective lines together, an *experimental* praxeology with strong explanation power. Contrary to the legend about the "reduction" ratio presented before, here the "reduction" ratio is greater than 1! However, this is only an experimental proof. Figure 13.7 (right) shows the front and left faces of a unit cube in a cavalier perspective, but also the image of the front face by a stretch by factor l and the image of this rectangle by a shear of angle α, which is precisely the left face of the cube; k being the "reduction" ratio. In this case, k is not equal to $\cos\alpha$, contrary to what the legend says. What does $\cos\alpha$ represent? It can be shown that $k\cos\alpha = l$. Hence, $\cos\alpha$ is the ratio of the factor of the stretch to the "reduction" ratio.

If we go on with the inquiry, we can find an application of affine transformations of the plane (especially shears and stretches) to relativistic (Galilean and Einsteinian) kinematics on the line (Pressiat, 2017, pp. 714–718). This application, due to Yaglom, echoes the fact that affine spaces have been defined by Weyl (1952) to build a mathematical theory of Einstein's general theory of relativity[1].

2.3. Concluding remarks

The starting point of the inquiry presented here was certain difficulties or gaps found by the French teaching profession regarding some geometrical objects at secondary school level. It illustrates that taking these difficulties seriously, as the starting point of a real study, leads to a rediscovery, or better understanding, of the raisons d'être and advantages of affine geometry, and to observing what existed and what disappeared from undergraduate mathematics education before the period of modern mathematics. It can be shown that, even at that time, it was possible to teach geometry in a very different way from the current set-up in France, mainly because there was the possibility of connecting it to other sectors of geometry such as Euclidean geometry of the plane and space. Another discovery (or rediscovery) is the link affine geometry has with the parallel projection of one plane on another, and therefore with the cavalier perspective.

This journey through recent mathematical and cultural time and space illustrates the possibility of providing the teaching profession with tools and ideas to feed the practical, experimental and even theoretical dimensions of the mathematical activities of teachers and students on these objects. It also shows that the profession in France suffers today from a culture dominated for several generations by the overvaluation of the formal dimension of mathematical organisations, at the expense of the practical, experimental and theoretical ones.

3 Interdisciplinary teaching as a problem of the teaching profession

Following the international rise of a focus on STEM education, the noosphere in Denmark has pushed educational initiatives in the same direction. The push has resulted in several curricular changes for primary and lower secondary schooling. The existing curriculum is at present still divided into separate disciplines named biology, geography, physics/chemistry at the lower secondary level and nature/technology at primary level; mathematics continues to be one separate entity at both levels. However, biology, geography and physics/chemistry were reformulated in 2014 with several "common competencies" applicable to all three school disciplines: "Inquiry competence", "Modelling competence", "Perspective competence" and "Communicative competence". This signified to everybody how much the three disciplines have in common and why it makes sense to take an integrated approach to teaching them. The curriculum further prescribes that pupils must work with at least four of six shared interdisciplinary topics (see list below).

Interdisciplinary topics for lower secondary science education

1. Production with sustainable use of natural resources
2. Sustainable energy supply – locally and globally
3. Drinkwater supply for future generations
4. Individual and societal discharge of substances
5. Impact of radiation on conditions for living organisms
6. Significance of technology for human health and living conditions.

This forces teachers to work, at least some of the time, across the disciplines, during the teaching year. In 2017, the incentive to teach the science disciplines in an integrated fashion was upped by the ministerial decision to instigate a joint examination, which had already been experimented with during 2016. This joint examination of the school science disciplines is now part of the obligatory final exams upon completing compulsory lower secondary school (Sillasen & Linderoth, 2017). Paradoxically, this development of requirements for lower secondary school teaching and examinations was not mirrored by similar official changes in the curriculum for teacher education in general. Of course, teacher educators individually have sought to take stock of the situation and attempted to prepare their students despite the disciplinary confinement.

The above-described development coincided with the start-up of the Advanced Science Teacher Education (ASTE) (Goldbech et al., 2018; Rasmussen

& Goldbech, 2013) project, a specialised version of the ordinary teacher education programme at University College Copenhagen, with the explicit aim of
teaching parts of the different science discipline curricula in integrated courses.
This has been fortuitous, and although it cannot be said that one of the developments precipitated the other in any linear way, it provides the backdrop for
many complex interactions within and across the institutional ecologies of "lower
secondary school" and "teacher education". In particular, it is noteworthy that
many of the conceptual words that figure prominently in Table 13.1 (e.g. "sustainable") can be identified as interdisciplinary rallying points in the genesis of
ASTE (Rasmussen, 2017).

The scope of this chapter does not allow us to go into the details of ASTE itself.
Our purpose is to convey how the ATD was able to elucidate the processes of
designing and implementing interdisciplinary teaching in teacher education. When
envisioning integrated teaching where there is none before, it is evident from the
ATD viewpoint that you initiate a process of didactic transposition (Bosch &
Gascón, 2006; Chevallard, 1985, 1989). Early in the design phase of ASTE, we
conducted an interview study among the teacher educators involved in the project
to uncover their reasons and arguments for how the disciplinary integration should
take place (Rasmussen & Winsløw, 2013). In the subsequent analysis using two
scales of didactic codeterminacy, it was possible to discern clear patterns of how the
involved disciplines interacted. The use of multiple scales to get a grip on interdisciplinarity was further explored in Rasmussen (2017), looking at the phenomenon across the different layers in the educational system: lower secondary school,
teacher education colleges, and universities.

Looking more closely at the design of an actual integrated mathematics and
biology course for pre-service teachers, Klaus Rasmussen (2015) explored the nature
of interdisciplinary knowledge for teacher educators, conceptualised as praxeological
organisations; respectively, organisations of declared knowledge to be taught and
the associated didactic organisations by the teacher educator (Chevallard, 1989;
Huillet, 2009). A perhaps discouraging finding was that integrated knowledge to be
taught resulted from a process where mono-disciplinary concerns took the fore,
making the topics for bi-disciplinary integration somewhat arbitrary. In a way, it
seems that the integration of disciplines was used to teach each of the disciplines
better, but without modifying the internal logic of traditional disciplines.

The continued subordination of interdisciplinarity to traditional disciplines is
reflected in the observations by Chevallard (2015) that there indeed appears to be
two paradigms at work in education: the one of "visiting monuments" and the one
of "questioning the world". Meaningful disciplinary integration appears to have a
greater chance of success in "questioning the world". However, it is not impossible
that promising integrated teaching can be experimented with in teacher education
even though the "monuments paradigm" is (still) the order of the day. Study and
research paths (Barquero, Bosch & Romo, 2015; Chevallard, 2009b; Winsløw,
Matheron & Mercier, 2013) were used in the design research of Rasmussen (2016),
where an interdisciplinary SRP was explored to scrutinise how this tool for didactic

engineering can be utilised to let pre-service teachers combine praxeologies from different disciplines while also considering associated didactic practices. When analysing the students logbooks using "tree diagrams" of questions and answers (Hansen & Winsløw, 2010), it became apparent that some students saw the integration between disciplines and associated didactic knowledge exclusively within one discipline while others showed signs of integration between the two disciplines without didactic consideration. There was scant evidence of students being able to combine both disciplines *and* didactic praxeologies. This just underlines the continued challenge for teacher education to provide settings where coming teachers can develop interdisciplinary praxeologies.

Coming back to the imposition of joint examinations in lower secondary school, it is worth noting that the ASTE project was not allowed to hold joint or alternative exams for their pre-service teachers. This meant that ASTE teachers underwent the ordinary mono-disciplinary evaluation, where there is no particular focus on, or indeed credit given for, competencies to address teaching and learning in interdisciplinary settings. Despite this fact, these teachers performed at least as well as their peers who did not attend the special ASTE programme.

4 The articulation between theoretical and practical knowledge as a problem of the teaching profession

Mathematics teacher knowledge consists of a complex combination of knowledge from theory and practice respectively. In teacher education, establishing coherence between theory and practice is a central and well-described problem (Bergsten *et al.*, 2009). However, developing competent mathematics teachers is not only a teacher education problem. When student teachers finish teacher education and become in-service teachers, a new problem arises. Most countries do not have programmes to further develop the recently qualified teacher (Winsløw *et al.*, 2009). Jeppe Skott (2001) describes how these teachers feel isolated and experience a conflict between what they have learned at the university and practice in schools, and James Stigler & James Hierbert (1999) see the teachers' isolation as a major obstacle to the development of mathematics teaching.

Developing mathematics teacher knowledge—for in-service as well as for pre-service student teachers—can be regarded as a theory-practice problem. In this perspective, the in-service teacher problem in many ways appears as "complementary" to the problem for pre-service student teachers. In-service teachers—especially the recently qualified—often feel isolated in the classroom and find it challenging to apply their theoretical knowledge in the daily practice, while pre-service student teachers usually study together in groups with fellow students, even if they often lack coherence with practice and relevant contributions from experienced practitioners. In other words, in-service teachers get almost all their experience from practice without relating this practice to theory and therefore have to develop their teacher knowledge from practice, while pre-service student teachers get almost all their experience from theory without much relationship to practice and therefore have to develop their teacher knowledge from theory only.

In the following section we will analyse data from a lesson study project with eight groups each with two in-service teachers and three or four pre-service student teachers. The main purpose of the project was to develop mathematics teacher education by establishing a milieu where student teachers experience a higher degree of coherence between the knowledge learned at the university college and the knowledge learned in practicum. We put two in-service teachers in each group to try to ensure that the dialogue between in-service teachers and student teachers did not end up like the (at least in Denmark) traditional dialogue where in-service teachers guide the student teachers. However, a very interesting side effect showed up to be that in-service teachers experienced collaboration with student teachers as very instructive.

The lesson study had two focus points: first, a didactical focus point, inquiry-based learning; and, second, a mathematical focus point, trigonometry. Both focus points are challenging for all participants. "Inquiry" is an important and often used word in the Danish curriculum for primary and lower secondary school, but still, mathematics teachers find it very demanding. Trigonometry is a new topic in the Danish curriculum for lower secondary school, and thus most teachers have little or no experience of teaching it.

Data consist of video recordings of the eight lessons and interviews with eight in-service teachers and eight student teachers—one from each group in the project. In the analysis we will focus on the participants' acquisition of teacher knowledge using notions from ATD like didactic transposition (Chevallard, 1985) and mathematical and didactic praxeologies (Chevallard, 1999).

4.1 How ATD deals with collaboration at the institutional level

The theory-practice problem in teacher education can be described as a matter of establishing coherence between theoretical knowledge learned at the university and practice knowledge learned in practicum. In other words, knowledge learned "outside" the school and knowledge learned "inside" the school. This distinction is also essential in what in ATD is called the *process of didactic transposition*: i.e. *scholarly knowledge* (SK) and *knowledge to be taught* (KTBT) outside the school and *taught knowledge* (TK) and *learned knowledge* (LK) inside the school. The theoretical education at university mainly considers SK and KTBT while practicum mainly refers to TK and LK.

The *internal didactical transposition* can be—and usually is—studied in practicum, but studies of what happens at the *external didactic transposition* are often "forgotten" between the theoretical studies at universities and practicum at schools (Østergaard, 2017). However, closer theoretical analyses show that there are two further essential theory-practice-relations in mathematics teacher knowledge and hence mathematics teacher education (Østergaard, 2015). First, there is a mathematical theory-practice relation between scholarly mathematical knowledge produced mainly at universities and *school mathematical knowledge* as described in KTBT and taught in TK. Second, there is a theory-practice relation between the practice block and the theory block in both the mathematical and the didactical praxeology, for instance in TK.

4.2 The practice block and the theory block

Both teachers and student teachers consider the practice block to be dominant in mathematical teaching in Denmark (Østergaard, 2016). The in-service teachers express that they are displeased with this, but they find it difficult to change. First, because the textbooks emphasise the practice block and second because they find it difficult to involve the theory block both for teachers and for pupils. Nevertheless, all interviewed participants in the project found it important to include both the practice block and the theory block in mathematical teaching and learning in schools and it is noteworthy that all participants assessed their study lesson as having an equal emphasis on practice and theory.

When it comes to teacher education, student teachers consider the theory block to be dominant. Therefore, it is more evident for student teachers to involve the theory block in their teaching in schools. The teachers' and the student teachers' two different foci might be at least one reason why TK in all study lessons in the project involves both the practice block and the theory block.

The following example is from a study lesson in 7th grade. The taught knowledge can be described by this mathematical praxeology:

T: Are these two triangles similar?
τ: Measure all angles in the two triangles. If the angles equal each other in pairs, the two triangles are similar.
θ: Two triangles in which angles in pairs equal each other are similar.
Θ: Definition of similarity.

The first part of the lesson considers the theory block. The teacher defines similarity from an example with two boxes of Toblerone chocolate and deduces the theorem (technology) about angles in similar triangles from a small exercise and a class discussion. In the next part of the lesson, the pupils are supposed to develop a technique to solve the task and link the practice block and the theory block together. The pupils get four cardboard triangles. The task is to investigate if any of the triangles are similar (two of them are similar) and, if they are, to explain why. The majority of the pupils put the triangles on top of each other to check whether the angles were the same in pairs. In the following class discussion, it turned out to be difficult for the pupils to combine their experience from the exercise with the definition and the theorem from the first part of the lesson, even though they had no problem solving the task. A few pupils succeeded, but most of them seemed to miss this point. This study lesson—and the other study lessons in the project—shows how difficult it is for pupils to establish coherence between the two blocks. All lessons involve both blocks, but only a few lessons succeeded in establishing coherence between the two blocks (as in this example).

4.3. Scholarly mathematical knowledge and school mathematical knowledge

A major aim of the lesson study project was to create an equal collaboration between in-service teachers and student teachers. All the interviewed participants afterwards expressed that this aim was achieved. For instance, an in-service teacher said:

> Maria: I think this dialogue was very different (from the usual guiding dialogue) because we were equal. I see myself as the more experienced in the classroom with the pupils but the student teachers are more up-to-date about theoretical issues. I teach grade 1 to 6 so trigonometry—sine and cosine—I do not quite remember. I haven't used these notions in my work so they are rather remote. In this case, the student teachers are more competent than I am.
>
> *(Østergaard, 2016, p. 248, own translation)*

Although the participants experienced collaboration as being equal, they still emphasised the two groups' different qualifications. Of course, they use different words, but most of them use terms like theoretical and practical knowledge:

> Michael: They (the student teachers) are "brisker" and more theoretical oriented. When you have been a teacher for some years, the theoretical knowledge fades into the background. Teaching becomes more practical although I hope there is some theory somewhere in the background. You just do not think about it.
>
> *(Østergaard, 2016, p. 251, own translation)*

The participants agree on the division in theoretical and practical qualifications in general, but they emphasise different theory-practice relations. Maria talks about the mathematical theory-practice relation between scholarly mathematical knowledge and school mathematical knowledge. Teachers do not deal with SK in their daily work and therefore have to take KTBT as starting point when a new topic is added to the curriculum. Several teachers indicated that their theoretical knowledge was insufficient, but it is noteworthy that none of them found it embarrassing. In their daily work, they compensate with experience and practice knowledge, but they found it instructive and developmental to work with student teachers who have the theoretical knowledge fresh in their memory.

5 Conclusions

The three cases outlined in the sections above are slightly different and also refer to different societal settings, in France and Denmark. However, they share a common concern about teacher education from the perspective of the profession and the problems it faces, which is a distinctive characteristic of how research about teacher education is conceived and conducted within the ATD.

From each case we can extract some interesting consequences. Thus, in the first case, about praxeologies for teaching as a problem of the teaching profession, the collection of praxeologies highlighted within it constitute an example of knowledge essential for the *identification* of mathematical praxeologies to be taught (Chevallard & Cirade, 2010). As Chevallard says, "scholarly mathematics knowledge is a function, not a substance" (Chevallard, 2007, p. 12, our translation). This travel in the professional and mathematical time and space shows that scholarly mathematics knowledge is not the same in all countries. It has great variability in its experimental and theoretical components and in the levels of explanation of its "raisons d'être". This travel should also provide the profession with infrastructural elements of didactic and mathematical praxeologies to cover their practical, experimental and even theoretical dimensions. Some of these elements proved to be fragile; their survival is at the mercy of the divisions in domains and sectors imposed by curricular programmes in France, the evolution of which has consequences that do not immediately appear. Ensuring their survival is part of the profession's responsibility.

The second case, about interdisciplinary teaching as a problem of the teaching profession, analyses a problem of the profession in Denmark that could certainly be extrapolated elsewhere. Forces in the Danish *noosphere* are pushing for a unified science (or STEM) discipline in primary and secondary education (Bohm et al., 2017; Nielsen & Horst, 2017), and lessons from the ASTE project tells us this needs to be reflected in teacher education. However, the traditional disciplines are strongly conditioned—institutionally and epistemologically—by higher education. Teacher educators themselves are primarily specialised into the various disciplines and their didactic. A weakening of content knowledge is feared, but a project like ASTE shows this to be unfounded. The theoretical notions of the ATD are well suited to keep a balanced view of disciplinary and didactic praxeologies when making a transition to an interdisciplinary curriculum and educational configuration that will be suited to meeting the needs of teachers facing an interdisciplinary future.

Finally, the third case, about the articulation of theoretical and practical knowledge as a problem of the teaching profession, evinces that creating coherence between theoretical and practical teacher knowledge is crucial for both student teachers and in-service teachers' further development of their teacher knowledge. The preceding analysis gives a brief idea of how collaboration between the two groups can create valuable new knowledge for both. In general, the project shows some interesting potential to develop mathematics teacher knowledge and mathematics teaching in general.

Beyond their differences, these three pieces of research clearly show to what extent the ATD is productive in inquiry about issues and challenges relating to the teaching profession.

Note

1 We appreciate that some of the descriptions above may be difficult for some readers to follow (full details are not possible given the space restrictions).

References

Barquero, B., Bosch, M., & Romo, A. (2015). A study and research path on mathematical modelling for teacher education. In K. Krainer & N. Vondrová (eds), *CERME 9 – Ninth Congress of the European Society for Research in Mathematics Education* (pp. 809–815). Prague: Charles University in Prague, Faculty of Education and ERME.

Bergsten, C., Grevholm, B., Favilli, F., Bednarz, N., Proulx, J., Mewborn, D., … Tsamir, P. (2009). Learning to teach mathematics: Expanding the role of practicum as an integrated part of a teacher education programme. In R. Even & D. L. Ball (eds), *The professional education and development of teachers of mathematics* (pp. 57–70). Boston, MA: Springer.

Bohm, M., Salomonsen, D., Quistgaard, N., Binau, C. F., Wøhlk, E. B., Jensen, L. V., & Kronvald, O. (2017). *Sammen om naturvidenskab: Anbefalinger til en national strategi for de naturvidenskabelige fag.* Copenhagen: Astra.

Bosch, M., & Gascón, J. (2006). Twenty-five years of the didactic transposition. *ICMI Bulletin*, 58, 51–65.

Bouligand, G. (1951). *L'accès aux principes de la géométrie euclidienne. Introduction à l'axiomatique du plan.* Paris: Librairie Vuibert.

Brannan, D. A., Esplen, M. F., & Gray, J. J. (1999). *Geometry*. Cambridge: Cambridge University Press.

Bybee, R. W. (2010). Advancing STEM education: A 2020 vision. *Technology and Engineering Teacher*, 70(1), 30–35.

Chevallard, Y. (1985). *La transposition didactique: Du savoir savant au savoir enseigné.* Grenoble, France: La Pensée Sauvage.

Chevallard, Y. (1998). Sur l'inadéquation de la formation première des professeurs de mathématiques de l'enseignement secondaire français. http://yves.chevallard.free.fr/spip/spip/article.php3?id_article=31.

Chevallard, Y. (1989). On didactic transposition theory: Some introductory notes. Paper presented at the International Symposium on Research and Development in Mathematics Education, Brastislava. http://yves.chevallard.free.fr/spip/spip/IMG/pdf/On_Didactic_Transposition_Theory.pdf.

Chevallard, Y. (1999). L'analyse des pratiques enseignantes en theorie anthropologique du didactique. *Recherches en Didactique des Mathématiques*, 19(2), 221–266.

Chevallard, Y. (2002). Organiser l'étude. 3. Écologie & regulation. In J.-L. Dorier (ed.), *Actes de la 11e École d'Été de Didactique des Mathématiques* (pp. 41–56). Grenoble, France: La Pensée Sauvage.

Chevallard, Y. (2007). Passé et présent de la théorie anthropologique du didactique. In L. Ruiz Higueras, A. Estepa & F. J. García (eds), *Sociedad, escuela y matemáticas: Aportaciones de la teoría antropológica de lo didáctico* (pp. 705–746). Jaén, Spain: Universidad de Jaén.

Chevallard, Y. (2009a). *La TAD face ou professeur de mathématiques*. Toulouse, France: Séminaire DiDiST.

Chevallard, Y. (2009b). Remarques sur la notion d'infrastructure didactique et sur le rôle des PER. Lecture given at the Journées Ampère, Lyon, France, May. http://yves.chevallard. free.fr/spip/spip/IMG/pdf/ Infrastructure_didactique_PER.pdf.

Chevallard, Y. (2015). Teaching mathematics in tomorrow's society: A case for an oncoming counter paradigm. In S. J. Cho (ed.), *The Proceedings of the 12th International Congress on Mathematical Education* (pp. 173–187). Cham: Springer International Publishing.

Chevallard, Y., & Cirade, G. (2010). Les ressources manquantes comme problème professionnel. In Ghislaine Gueudet & Luc Trouche (eds), *Ressources vives: Le travail documentaire des professeurs en mathématiques* (pp. 41–55). Rennes, France: Presses Universitaires de Rennes et INRP.

Cirade, G. (2006). Devenir professeur de mathématiques : Entre problèmes de la profession et formation en IUFM : Les mathématiques comme problème professionnel. Université de Provence Aix-Marseille I, France.

Deltheil, R., & Caire, D. (1944). *Géométrie, classe de mathématiques et préparation aux grandes écoles*. Paris: J.-B. Baillière et Fils.

Goldbech, O., Aarby, J., Rasmussen, K., Winsløw, C., & Østergaard, C. H. (2018). *ASTE: Rapport til Lundbeckfonden*. Københavns Professionshøjskole, Denmark.

Hansen, B., & Winsløw, C. (2010). Research and study course diagrams as an analytic tool: The case of bi-disciplinary projects combining mathematics and history. In M. Bosch (ed.), *Un panorama de la TAD* (pp. 257–263). Bellaterra, Spain: Centre de Recerca Matemàtica.

Huillet, D. (2009). Mathematics for teaching: An anthropological approach and its use in teacher training. *For the Learning of Mathematics*, 29(3), 4–10.

Köhler, J., Höwelmann, R., & Krämer, H. (1973). *Analytische Geometrie und Abbildungsgeometrie in vektorieller Darstellung*. Frankfurt, Germany: Diesterweg.

MEN (Ministère de l'Éducation Nationale) (2012). *Mathématiques Série STD2A: Perspectives cavalières, parallèles et créations graphiques*. Collection Ressources pour la classe de première générale et technologique. Paris: Eduscol.

Mercier, D.-J. (2012) *Oral 1 du CAPES Mathématiques : Plans et approfondissements de cinq leçons de la liste 2013*. Paris: Publibook.

Nielsen, K., & Horst, S. (2017). Sammen om naturfagsstrategien. *MONA – Matematik- og Naturfagsdidaktik*, 3, 61–71.

Østergaard, K. (2015). A model of theory-practice relations in mathematics teacher education. In K. Krainer & N. Vondrová (eds), *Proceedings of the Ninth Conference of the European Society for Research in Mathematics Education* (CERME9, 4–8 February) (pp. 2888–2894). Prague: Charles University, Faculty of Education and ERME.

Østergaard, K. (2016). Teori-praksis-problematikken i matematiklæreruddannelsen. Roskilde University, Denmark.

Østergaard, K. (2017). Theory and practice in mathematics teacher education. In G. Cirade et al. (eds), *Évolutions contemporaines du rapport aux mathématiques et aux autres savoirs à l'école et dans la société* (pp. 899–918). Toulouse, France: IUFM de Toulouse.

Parzysz, B. (1989). Représentations planes et enseignement de la géométrie de l'espace au lycée : Contribution à l'étude de la relation voir/savoir (Doctoral dissertation). Université Paris 7, France.

Pressiat, A. (2017). Éléments sur les apports de la théorie anthropologique du didactique à la profession et leur réception. In G. Cirade et al. (eds), *Évolutions contemporaines du rapport aux mathématiques et aux autres savoirs à l'école et dans la société* (pp. 679–725). Toulouse, France: IUFM de Toulouse.

Rasmussen, K. (2015). Didactic transposition of mathematics and biology into a course for pre-service teachers: A case study of 'Health: Risk or Chance?'. In M. Achiam & C. Winsløw (eds), *The relationships and disconnections between research and education* (vol. 39) (pp. 107–120). Copenhagen: University of Copenhagen.

Rasmussen, K. (2016). The direction and autonomy of interdisciplinary study and research paths in teacher education. *Journal of Research in Mathematics Education*, 2, 158–179.

Rasmussen, K. (2017). A layered model of didactic codetermination in science teacher education: Institutional conditions and constraints when planning multidisciplinary teaching of energy topics. In G. Cirade *et al.* (eds), *Évolutions contemporaines du rapport aux mathématiques et aux autres savoirs à l'école et dans la société* (pp. 919–939). Toulouse, France: IUFM de Toulouse.

Rasmussen, K. (2017). The emergence and institutional co-determination of sustainability as a teaching topic in interdisciplinary science teacher education. *Environmental Education Research*, 23(3), 348–364.

Rasmussen, K., & Goldbech, O. (2013). ASTE-profil i den nye læreruddannelse. *MONA – Matematik- og Naturfagsdidaktik*, 4, 92–95.

Rasmussen, K., & Winsløw, C. (2013). Didactic codetermination in the creation of an Iitegrated math and science teacher education: The case of mathematics and geography. In B. Ubuz, Ç. Haser & M. A. Mariotti (eds), *Proceedings of CERME 8* (pp. 3206–3216). Ankara: Middle East Technical University.

Reinhardt, F., & Soeder, H. (1999). *Atlas des mathématiques*. Paris: LGF – Le livre de poche.

Sillasen, M. K., & Linderoth, U. H. (2017). Tværfaglig undervisning i folkeskolens naturfag. *MONA – Matematik- og Naturfagsdidaktik*, 3, 19–38.

Skott, J. (2001). The emerging practices of a novice teacher: The roles of his school mathematics images. *Journal of Mathematics Teacher Education*, 4(1), 3–28.

Stigler, J. W., & Hierbert, J. (1999). *The teaching gap: Best ideas from the world's teachers for improving in the classroom*. New York: The Free Press.

Weyl, H. (1952). *Space-time-matter*. New York: Dover Publications.

Winsløw, C., Bergsten, C., Butlen, D., David, M., Gómez, P., Grevholm, B., … Wood, T. (2009). Chapter 1.2.3. In R. Even & D. L. Ball (eds), *The professional education and development of teachers of mathematics*. New ICMI Study Series, vol. 11 (pp. 93–101). Boston, MA: Springer

Winsløw, C., Matheron, Y., & Mercier, A. (2013). Study and research courses as an epistemological model for didactics. *Educational Studies in Mathematics*, 83(2), 267–284.

Yaglom, I. M., & Shenitzer, A. (1973). *Geometric transformations III*. Washington, DC: The Mathematical Association of America.

INDEX